Advance Praise for

AMPHIBIANS AND REPTILES
OF LA SELVA, COSTA RICA,
AND THE CARIBBEAN SLOPE

AMPHIBIANS AND REPTILES OF LA SELVA, COSTA RICA, AND THE CARIBBEAN SLOPE

AMPHIBIANS AND REPTILES

OF LA SELVA, COSTA RICA, AND THE CARIBBEAN SLOPE: A COMPREHENSIVE GUIDE

Craig Guyer
Maureen A. Donnelly

UNIVERSITY OF CALIFORNIA PRESS

Berkeley   Los Angeles   London

University of California Press
Berkeley and Los Angeles, California

University of California Press, Ltd.
London, England

Library of Congress Cataloging-in-Publication Data

Guyer, Craig, 1952–.
    Amphibians and reptiles of La Selva, Costa Rica, and the Caribbean Slope :
    a comprehensive guide / Craig Guyer, Maureen A. Donnelly.
        p.   cm.
    Includes bibliographical references and index.
    ISBN 0–520-23758-7 (cloth : alk. paper) — ISBN 978-0-520-23759-9 (pbk. : alk.
    paper)
    1. Amphibians—Costa Rica—Heredia (Province)   2. Reptiles—Costa Rica—
Heredia (Province)   3. Estación Biológica La Selva (Costa Rica)   I. Donnelly,
Maureen A., 1954–   II. Title.

    QL656.C783G89 2004
    597.9'097286'4—dc22   2004013039

Manufactured in China
19   18   17   16   15   14   13   12   11
10   9   8   7   6   5   4   3   2

The paper used in this publication meets the minimum requirements of
ANSI/NISO Z39.48–1992 (R 1997) (Permanence of Paper). ♾

Cover: Strawberry Poison Frog, photograph by Paul Buttenhoff.

CONTENTS

Acknowledgments   vii

INTRODUCTION   **1**

Geography, Climate, and Vegetation   7

A History of Herpetological Research at La Selva   11

Diversity   14

Conservation   17

AMPHIBIA   **21**

Gymnophiona   24

Caudata   27

Anura   33

REPTILIA   **103**

Testudines   106

Squamata   117

Crocodylia   233

ADDITIONAL SPECIES   **239**

Field Data Sheet for Amphibians and Reptiles   251

Glossary   253

Literature Cited   265

Art Credits   285

Index   287

Plate sections follow pages 62 and 170

ACKNOWLEDGMENTS

Special thanks go to Jay M. Savage for starting and maintaining modern herpetological research at La Selva. He directly and indirectly provided opportunities for us to begin long-term research at the site. Many of the observations and data summarized here were recorded while we were supported, in various ways, by research funds from the Jessie Smith Noyes Foundation. Additionally, we have maintained this research effort by participating in courses sponsored by the Organization for Tropical Studies. The past station directors, Deborah and David Clark, Cynthia Echeverria and Bruce Young, and Robert Matlock, and the current director, Luis Diego Gomez, took special interest in the herpetofauna project and extended many courtesies to us during our stays at the station.

We benefited from the support of numerous station workers, students, and researchers, who aided us by catching animals, testing our keys, and recounting natural history observations. We do not have enough space to thank them all individually but acknowledge here that our efforts would have failed without them. However, Orlando Vargas, Michael and Patricia Fodgen, David and Deborah Clark, Manuel Santana, and Harry W. Greene deserve special thanks for long discussions about the herpetofauna and special efforts in catching specimens. Ryan Dibaudio, Tonya Haff, Sharon Hermann, Heidi Kloeppel, Andres Vega, and Lori Wollerman provided notable assistance in testing the keys. James Watling and Scott Boback kindly provided their data on the sizes of frogs from La Selva. For information on rare species and museum collections from La Selva and the surrounding area we thank Jonathan A. Campbell (UTA), Roy W. McDiarmid (USNM), Jay M. Savage (CRE), John E. Simmons (KU), David B. Wake (MVZ), and Thomas Trombone (AMNH). Also deserving special thanks is James B. Murphy for taking interest in the final product and providing a number of fine suggestions for its organization and content. Emmett L. Blankenship, Paul A. Buttenhoff, Pete N. Lahanas, Kirsten E. Nicholson, Robert N. Reed, and Jay M. Savage carefully read the manuscript and made many fine suggestions. The words that survived these reviews were converted to more appropriate grammar by the careful copyediting at Impressions Book and Journal Services. We, of course, accept responsibility for all shortcomings that remain. We thank Sharon M. Hermann and Steven F. Oberbauer for their steadfast support of "Los Beezers."

INTRODUCTION

BEFORE THE ADVENT of modern agriculture, the northeastern third of Costa Rica was covered by a lush and diverse type of vegetation referred to as lowland tropical wet forest. This habitat extended as a nearly continuous canopy along the Caribbean coast northward to Veracruz, Mexico, and southward to western Panama. Today it is one of the most biologically diverse forests in the world. Conversion of these lands to agricultural uses started in the 1940s, and the ancestral forest was fragmented into smaller and smaller units. Fortunately, the people of Costa Rica recognized the value of their rich biotic resources and set aside examples of this forest type, as well as other ecosystems, in a series of public reserves. Examples of Costa Rican lowland tropical wet forest are found in the Refugios Nacionales de Vida Silvestre Caño Negro, Barra del Colorado, and Gandoca-Manzanillo; the Parques Nacionales Cahuita and Tortuguero; and the Reserva Biologica Hitoy-Cerere.

In addition to these public reserves, the La Selva Biological Station, a private research station operated by the Organization for Tropical Studies (fig. 1), is well known as an important example of lowland tropical wet forest. Compared with the other reserves, La Selva is unimpressive in size. However, it is connected to Braulio Carrillo National Park by a protected zone of pristine and regenerated lands. This national park is part of the Cordillera Volcánica Central Biosphere Reserve, a large conservation area designed to protect the significant biological diversity of the Cordillera Central of Costa Rica. Thus, La Selva is a component of the lowland forest reserves of Costa Rica and contributes to an elevational transect of lands representing the rich forests of the Caribbean slopes of Central America. Additionally, this locality is a premiere research site for study of the life history of the flora and fauna of lowland tropical wet forests.

Amphibians and reptiles are indispensable inhabitants of the lowland tropical wet forests of Costa Rica. Although these creatures are less familiar to the public than charismatic groups such as mammals and birds, amphibians and reptiles are unusually abundant and represented by an exceptionally rich collection of species. Additionally, amphibians and reptiles occupy all major habitats (aquatic, subterranean, terrestrial, arboreal), exhibit diverse reproductive strategies (from egg laying to live birth), have complex behaviors (e.g., care of eggs and offspring by both parents), and display varied diets (from herbivory to carnivory). Some radiations of amphibians (e.g., frogs and toads) and reptiles (e.g., lizards and snakes) rival the mammalian radiation both in the numbers of species living today and the variety of ecological roles that they fill. For these reasons, information about amphibians and reptiles at La Selva is of value to scientists for continued expansion of our knowledge about these creatures, and to amateur naturalists to encourage understanding of this understudied component of the natural environment.

In this book we summarize our more than 40 combined years of research on amphibians and reptiles at La Selva. Our primary goal is to introduce researchers and amateur naturalists to key life history features of each species on the site. Another goal is to provide an identification guide to all species of amphibians and reptiles that occupy this region of Costa Rica. The book is divided into three sections. In the first section we describe the habitats of La Selva, the history of herpetological research there, general patterns of diversity, and efforts to conserve the animals in this area. In the other two sections, we divide information taxonomically, one section for the class Amphibia and another for the class Reptilia. We begin each taxonomic section with a description of the key characteristics of each class, followed by a summary of the life history of class members and general descriptions of their foraging habits and major predators. A key (see below) allows identification of the major orders within each class. We then present diagnostic features, life histories, and tips for finding and capturing members of each order. We end these sections with keys to the species within each order at La Selva. Descriptions and photos are provided for each species, organized by the family to which each species belongs. Each family is placed in phylogenetic order (Ford and Cannatella 1993 for frogs; Gaffney and Meylan 1988 for turtles; Estes et al. 1988 for lizards; Greene 1997 for snakes). The text for each family starts with a general description of the family and is followed by species accounts.

A significant feature of this book is the taxonomic keys—devices that present a series of paired choices, organized to allow identification of an unknown organism. At each step in a key, you compare the unidentified organism to two statements. One statement is true of the organism in question and the other is false. Associated with the statement that fits is a number that indicates the

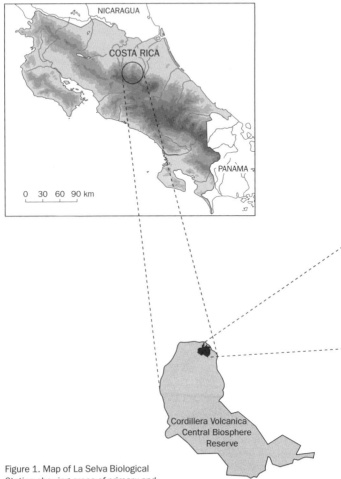

Figure 1. Map of La Selva Biological
Station showing areas of primary and
secondary forest and other habitat
types. Insets show the position of La
Selva within the Cordillera Volcánica
Central Biosphere Reserve, and the
location of the Biosphere Reserve
within Costa Rica.

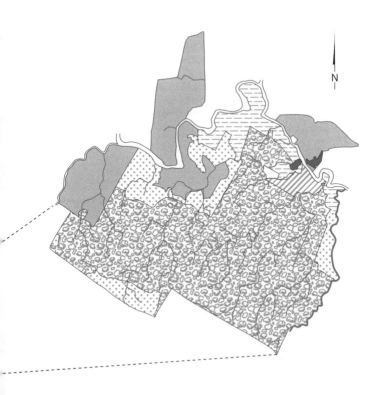

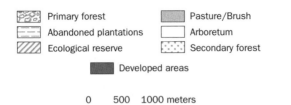

	Primary forest		Pasture/Brush
	Abandoned plantations		Arboretum
	Ecological reserve		Secondary forest
	Developed areas		

0   500   1000 meters

next paired statements to be evaluated. You repeat the process of comparing the organism to pairs of statements until a true statement provides the identity of the organism.

In this book, keys are based primarily on color patterns and are designed so that careful notes taken in the field can be used to identify individuals that escape capture or are released after observation. The keys should be used in conjunction with those of Savage and Villa (1986) and Savage (2002) for the herpetofauna of Costa Rica. These volumes are more authoritative but require extensive knowledge of features of external and internal anatomy. At the back of the book, we provide a data sheet to aid the recording of appropriate characteristics of animals seen but not captured, or captured and released in the field. A hand lens is needed to observe some of the traits used in our frog and lizard keys, and a measure or estimate of body size is helpful for identification.

For each species of amphibian and reptile known from La Selva, we provide an overview of its natural history. These descriptions summarize our field notes and experiences and are not meant to be exhaustive. Rather, we intend to indicate where each species is likely to be found and some brief details of its habits. These should aid in identification of problem taxa. Where possible we have included pertinent information regarding size (body length), geographic distribution, identifying features, sexual dimorphism, habitat, diet, and reproduction. Data regarding size are based on measurements taken from animals captured and released, or samples gathered during extensive collection of museum specimens at La Selva (Lieberman 1986). We use measures (in millimeters) of snout-to-vent length (SVL)—from the tip of the snout to the cloacal opening—for all frogs, salamanders, and lizards; carapace length (CL)—from the front of the shell to the back of shell—for all turtles; and total length (TL)—from the tip of the snout to the tip of the tail—for all caecilians, crocodilians, and snakes. Where possible, we provide lengths for the largest adult male and adult female separately; in those cases where sex cannot be determined readily, we give the largest size of all individuals. These data are from animals captured on La Selva property or from Braulio Carrillo National Park near its border with La Selva.

We also provide the most recent scientific names for taxa known from La Selva. Some long-familiar generic names have been replaced with nomenclature that reflects monophyletic lineages based on current systematic evidence. In such cases we include the old generic designations in the species accounts. We do not provide a complete review of synonymies; rather, we indicate names likely to have been used during the last 50 years. Our treatment of iguanians follows the family designations of Frost and Etheridge (1989), in which the old "Iguanidae" is divided into eight families.

For each species we include common names used by local people of the Sarapiquí region, as well as English common names. For the former we include names listed by Janzen (1983), and for the latter we list names used by Greene (1997), Hayes et al. (1986), Iverson (1992), Janzen (1983), Leenders (2001), and Liner (1994). A few other names, from unknown sources but used consistently at La Selva, are included. Finally, we provide a section that describes species known from the Caribbean lowlands of Costa Rica, but which are not known from La Selva. This information allows identification of species that may be present but have not yet been collected at La Selva and allows this book to be useful at other sites throughout the Caribbean lowland tropical wet forests of Costa Rica.

Geography, Climate, and Vegetation

La Selva is located in Heredia Province of northeastern Costa Rica, just south of the town of Puerto Viejo. It covers 1,536 ha of rolling topography bordered to the north by the Rio Sarapiquí, to the west by the Rio Péje (a tributary of the Rio Sarapiquí), and to the east by the Rio Puerto Viejo and a tributary of this river, the Quebrada Sabalo Esquina. The property is lowest in elevation at the northern end (35 m above sea level, where the Rios Sarapiquí and Puerto Viejo meet) and rises in elevation toward the southern end (137 m above sea level at its highest point), where the property joins the lower reaches of the Braulio Carrillo National Park. Therefore, all of the plants and animals at La Selva are adapted to tropical lowland wet forests.

About 2 million years ago lava flows from the volcanoes of the Cordillera Central of Costa Rica (Sollins et al. 1994) began forming the current topography. Along with lava, these volcanoes created massive areas of mudflow, some of which are found on La

Selva. Five major creeks drain the site, which, along with the boundary rivers, carry soil sediments that are occasionally deposited along the banks and flood plains during periods of intense rain. Such deposits have been created for millennia, so that younger (recent) layers occur along areas close to watercourses, and older (Pleistocene) layers are found farther upslope. Thus, La Selva contains upland areas largely of volcanic origin, lowland areas of recent alluvial deposits, and older alluvial terraces between these extremes.

La Selva is a wet place, averaging about 4 m of rain per year. Moisture-laden air moves toward the west off of the Caribbean Ocean and deposits this moisture as rain as the air rises over the Cordillera Central. Although rain occurs every month of the year at La Selva, amounts change seasonally. Most precipitation falls from May through December, a period referred to as the rainy season. However, even during this interval, rainfall tends to peak twice, once during June and July and once during November and December (Sanford et al. 1994). Between these peaks is a one-month period of limited rainfall, typically during September or October (often referred to as the little dry season, or *veranillo*). Rainfall is less during the months of January through April, a period referred to as the dry season. Although rainfall decreases during this time, it rarely stops, and so, even during the dry season, monthly rainfall totals typically exceed 100 mm.

The special combination of abundant rain, low elevation, and equatorial location (below the Tropic of Cancer) places La Selva within the lowland tropical wet forest climate zone (Hartshorn and Hammel 1994). This zone is characterized by warm daily mean temperatures throughout the year and temperatures that fluctuate more widely during a 24-hour period than they do seasonally. Tree growth within this climate zone is exceptional, creating a tall canopy that reaches up to 55 m above the forest floor at La Selva. Woody vines (lianas) grow up to the top of the canopy and can extend across several canopy trees, connecting these trees and providing avenues across which canopy organisms travel. Below the canopy multiple layers of trees and shrubs fill much of the space. At every vegetative level, epiphytic plant life can be found, a feature that creates habitats not available at less heavily vegetated localities. Approximately half of La Selva is covered with primary forest, meaning that the forest has not been cut by recent industrial operations. This does not mean that the forest has not been disturbed by human activities. In fact, for centuries native human cultures occupied La Selva and likely disturbed the forest by cutting trees and burning some areas (Sanford et al. 1994). Nevertheless, the forest reserves at this site are of exceptional quality and are representative of a vanishing forest ecosystem.

In describing the habitats present at La Selva, we follow the system of Hartshorn and Hammel (1994) and recognize five native vegetative types: primary forest on rolling terrain, primary forest on old terrace, primary swamp forest, open swamps, and riparian forest. In the four forest types the subcanopy legume gavilán *(Pentaclethra macroloba)* is dominant, accounting for approximately 40 percent of all trees. La Selva has a flora that is rich in the total number of tree species present, and the dominance of this one species is unusual for lowland tropical wet forest sites. Detailed explanations for this feature are lacking, but it is thought that some combination of tolerance to soil infertility, a relatively restricted dry season, and a lack of significant seed predators for gavilán has allowed this swamp-dwelling species to expand throughout all habitats at La Selva (Hartshorn 1983).

Two forest types, one on rolling terrain and the other on old terrace, are distinguished primarily by soil type; the former occurs on old lava flows, and the latter occurs on old alluvial deposits. These upland habitats are the most common forests at La Selva and cover about 55 percent of the property. A visually striking feature of these habitats is the canopy emergent trees, exceptionally tall species that rise above the crowns of most forest tree species. A characteristic canopy emergent of the forests on rolling terrain is ajillo *(Pithecellobium elegans)*, whereas almendro *(Dipteryx panamensis)* is a canopy emergent characteristic of old alluvial soils (Clark 1994). In both forest types the subcanopy palms palma conga *(Euterpe macrospadix)*, consuelo de mujer *(Iriartea deltoidea)*, maquenque *(Socrotea exorrhiza)*, and Costa Rica feather palm *(Welfia georgii)* are common, as are the trees mata gente *(Dendropanax arboreus)*, copal *(Protium pittieri)*, yaya *(Unonopsis pittieri)*, and lengua del diablo *(Warscewiczia coccinea)* (Hartshorn and Hammel 1994). The understory is dominated by short palms of the genera *Bactris, Geonoma,* and *Asterogyne,* and shrubs of the families Rubiaceae, Piperaceae, and Melastomataceae.

In areas of recent alluvial deposits, drainages can become

clogged, creating a third habitat called swamp forests—lands in which soils retain water for long periods of time each year. Foot travel through such areas is a challenge because the accumulated fine soils typically will not support the weight of an average person. Such areas have a distinctive flora in which Nicaraguan water tree *(Carapa nicaraguensis)*, guácimo colorado *(Luehea seemannii)*, bola de oro *(Otoba novogranatensis)*, huevos de burro *(Pachira aquatica)*, and sangregado *(Pterocarpus officinalis)* are relatively common canopy trees. In the subcanopy zone the tree species tabacón *(Grias cauliflora)* and ajillo are swamp specialists, as are the understory plants clavillo *(Adelia triloba)*, coyolillo *(Astrocaryum alatum)*, *Bactris longiseta*, Costa Rican fruta de pava *(Chione costaricensis)*, and cafecillo *(Psychotria chagrensis)* (Hartshorn and Hammel 1994). Swamp forests cover a relatively small area of La Selva (approximately 60 ha) and are distributed in patches of variable size.

A fourth vegetation type at La Selva is associated with open swamps. These are small, treeless areas that hold water for extended periods of time during the wet season. They are located in low-lying areas of recent or old alluvium where soils are poorly drained and, therefore, tend to retain rainwater or floodwaters they receive from major rivers and streams. The vegetation of such sites has a periphery dominated by the aroid anturto blanco *(Spathiphyllum freidrichsthallii)* and a center dominated by the grass zacate grande *(Panicum grande)* (Donnelly and Guyer 1994). Other than gaps formed by large fallen trees, open swamps are one of the few places in the forested areas of La Selva where direct sunlight penetrates to the forest floor. Such swamps are distinctive, and only a handful of them are present on the La Selva property.

The final primary vegetation type at La Selva is riparian forest. Because two major rivers border the property and several major creeks traverse it, riparian forest, the vegetation zone that lines these watercourses, is a significant component of the primary forests. The two most characteristic tree species of this habitat are chilamate *(Ficus insipida)* and sota caballo *(Pithecellobium longifolium)*, which grow along the banks of the rivers; other plant species that are frequently found in these vegetative zones are Ruiz's guabo *(Inga ruiziana)*, guácimo colorado, and turrú colorado *(Myrica splendens)* (Hartshorn and Hammel 1994). The banks of these areas frequently are remarkably steep and, seemingly, held together with tree roots. Additionally, water levels of the rivers and major streams change rapidly during intense rains, creating cavities along these banks that provide hiding places for many kinds of animals.

In addition to largely pristine habitats, La Selva has extensive areas influenced by recent human agricultural activities. The principle alteration to the landscape surrounding La Selva is clearings created for cattle ranching and subsistence farming. In addition, extensive areas have been cleared for commercial farming of bananas, oranges, and cacao (the source of chocolate). Lands on La Selva were altered for similar purposes (see below), largely along the flood plains of the two major rivers. Some of these sites remain as open pastures, whereas other areas have become secondary forests in various stages of reversion to the primary vegetation types described above. In general, areas of second growth can be distinguished from primary forest by the structure of the understory. In primary forests the understory is open and easy to see and walk through. In secondary forest, shrubs and vines are thick and seemingly impenetrable; it is difficult to walk off the trail systems in these areas.

A History of Herpetological Research at La Selva

The earliest collections of amphibians and reptiles from the region surrounding La Selva were made by Pablo Biolley, a teacher at the Colegio de San José during the late 1800s. Biolley was a prodigious naturalist who collected Alvarado's Salamander *(Bolitoglossa alvaradoi)* in 1891 from "Sarapiquí," and the Dry Forest Anole *(Norops cupreus)* in 1892 from "Puerto Viejo de Sarapiquí." Both specimens initially were deposited in the Museo Nacional de Costa Rica but eventually were included in an Edward D. Cope Collection acquired by the American Museum of Natural History (AMNH A11725 and AMNH A16353, respectively). Neither taxon has ever been collected at La Selva, but a specimen of Alvarado's Salamander from nearby Rio Frio is in a collection at the Florida State Museum. We have examined the specimen of the Dry Forest Anole and have confirmed that identification. Because this species is not known from the Atlantic versant of Costa Rica, we presume that the locality for this specimen is incorrect.

The modern herpetological exploration of the Sarapiquí region began in the late 1940s and early 1950s with the efforts of Edward H. Taylor, an influential herpetologist from the University of Kansas who was interested in documenting the herpetofauna of the entire country. Taylor (1951) first collected amphibians and reptiles from Costa Rica in the summer of 1947 on invitation from the Rector of the Universidad de Costa Rica. His subsequent expeditions took him to the Caribbean lowlands, where he was a guest of Dr. Leslie R. Holdridge during the summer of 1954 (Taylor 1955). Finca La Selva was purchased by Holdridge in 1953 (McDade and Hartshorn 1994), and he encouraged biological exploration of the finca and surrounding property. Surprisingly, only a single species, the Strawberry Poison Frog *(Dendrobates pumilio)* (but described as *D. typographicus*), was reported by Taylor (1958; KU 36523–43). This undoubtedly resulted from Holdridge's reticence in allowing collection on his property. The collection data for the series of Strawberry Poison Frog specimens collected by Taylor list the locality as "Puerto Viejo N. Limon Prov." We assume this series was collected on or near Holdridge's Finca la Selva because Puerto Viejo de Limón is in the center of Limón Province, and Puerto Viejo de Sarapiquí, although in Heredia Province, is near the northern part of Limón Province. Taylor (1958) mentioned that he was a guest of the Holdridges during this time.

In the 1960s two other influential herpetologists from Kansas, William E. Duellman and Henry S. Fitch, visited Puerto Viejo de Sarapiquí. Duellman studied the hylid frogs in the region (Duellman 1967a, 1967b, 1970), but he did not make collections at La Selva. This research culminated in the publication of a monograph on the hylid frogs of Middle America (Duellman 1970). Fitch visited La Selva first in 1965 as part of an Organization for Tropical Studies (OTS) field course (Fitch 1975) and returned to La Selva during 1972–1973 (Fitch 1976). His research at La Selva and in nearby areas focused largely on the life history of lizards (Fitch 1973a, 1973b, 1975, 1976, 1982; Fitch et al. 1976) and uncovered patterns that became the focus of later demographic studies (see below). Collections made by Taylor, Duellman, and Fitch are the foundation of the museum materials from the Sarapiquí of Costa Rica housed at the Museum of Natural History at the University of Kansas.

The most comprehensive exploration of La Selva's herpeto-fauna began with Jay M. Savage, who, in the late 1950s, started a long-term study of Costa Rica's herpetofauna. The first of Savage's many field crews arrived in Costa Rica in 1959 for a collecting expedition. This crew consisted of Arden H. Brame, Jr., Arnold G. Kluge, and Robert J. Lavenberg. They were invited to visit La Selva and obtained the first extensive collections from Costa Rica that are now housed at the Los Angeles County Museum in the Costa Rica Expeditions collections. Savage first visited La Selva in August of 1961, an event that marked the beginning of a long period of investigation. Eventually, Holdridge tired of attempting to create a viable farm and sold the property to OTS (McDade and Hartshorn 1994), and the site became a mainstay for graduate courses in tropical ecology. For many years one of Savage's students, Norman J. Scott, Jr., served as an OTS instructor and collected a long-term series of leaf litter samples at La Selva and other sites as part of an extended field study; these data were summarized in Scott (1976). Additionally, Scott's experience at the site was instrumental in the development of the first published herpetofaunal list for La Selva (Scott et al. 1983). The incredible diversity and often fantastic abundance of amphibians and reptiles in Scott's leaf litter samples led Savage and Ian Straughan to develop an expanded project to describe the community dynamics of leaf litter amphibians and reptiles. This research accumulated an extensive sample of La Selva's herpeto-fauna, summarized in Lieberman (1986).

With the conversion of La Selva into a full-time research station, and especially with the success of OTS in providing research opportunities for graduate students (Clark 1990), herpetological research at the site has grown. As with the study of any site, this growth occurred because there was background information documenting the species present and at least some accumulated natural history data clarifying the phylogenetic relationships of these species and demonstrating important ecological and evolutionary interactions. From this foundation, recent herpetological research has added behavioral studies of communication (Strieby 1998; Wollerman 1995), reproduction (Donnelly and Guyer 1994; Roberts 1994a, 1994b; Watling and Donnelly 2002), territoriality (Robakiewicz 1992), activity (Winter 1987), and foraging (Greene and Santana 1983); experimental studies of population and community ecology (Donnelly 1989a, 1989b; Guyer 1988a, 1988b; Talbot 1977, 1979); comparative studies of physiology (van Berkum 1986, 1988) and predation (Brodie 1993); and further community studies of the leaf litter herpetofauna (Fauth et al. 1989; Heinen 1992; Slowinski et al. 1987).

Diversity

The herpetofauna of La Selva contains 138 known species. This remarkably rich assemblage is comparable to that of other Central American sites (Donnelly 1994a, 1994b; Guyer 1994a, 1994b). Many of the samples from La Selva come from the 587 ha of the original Holdridge property, with fewer samples from the 949 ha of the newer additions (see McDade and Hartshorn 1994). This relatively small sampling area confounds species-diversity comparisons between La Selva and other Neotropical sites with larger area (e.g., the 1,532,000 ha of Manu National Park, Peru; Terborgh 1990) or composite sampling areas (e.g., Iquitos Region, Peru; Rodriguez and Duellman 1994).

The diverse herpetofauna is a result of and maintained by several factors related to geological history, forest structure, and climate. The fossil record of many of the subfamilies and genera of amphibians and reptiles at La Selva can be traced back in time to the Eocene (Savage 1982). Thus, the evolutionary history of these groups was affected by the complex geological history of Central America. Accumulating evidence suggests that the intricate interactions of the Cocos, Nazca, North American, South American, and Caribbean plates created at least two island arcs, or land bridge connections, between North and South America (Guyer and Savage 1986; Rosen 1976, 1985; Savage 1982). Additionally, patterns of sea level invasion and subsidence (Savage 1966; Stuart 1966) resulted in the separation of North and Central American taxa across the Isthmus of Tehuantepec. With the reconnection of North and South America via the closing of the Panamanian Portal in Miocene/Pliocene times (Guyer and Savage 1986; Savage 1982) and the final drying of the Isthmus of Tehuantepec in the late Pliocene (Savage 1966; Stuart 1966), mixing of the isolated elements of the herpetofaunas occurred. Part of La Selva's diversity results from the fact that it contains taxa that arose from a long period of independent evolution in Central America, taxa that dispersed from centers of evolution in South America, and taxa that dispersed from young and old centers of evolution in North America (Savage 1982).

The dynamics of the forest structure also contribute to herpetofaunal species richness at La Selva. Although there are no known altitudinal migrants—as there are for birds, butterflies, and moths (Blake et al. 1990; Janzen 1983; Levey and Stiles 1994)—a small number of amphibian and reptilian species are associated with gaps. The Central American Whiptail *(Ameiva festiva)*, a lizard commonly found on trails, is active at high body temperatures that it reaches by basking in open sunlight (Echternacht 1983; van Berkum 1986). Before the advent of trails and pastures, this species may have maintained its populations by colonizing areas opened to sunlight by tree falls or by colonizing similarly disturbed sites along rivers. Additionally, the Gray-eyed Leaf Frog *(Agalychnis calcarifer)* has only been observed to reproduce in pools of water on the trunks of fallen canopy trees (Marquis et al. 1986; Roberts 1994a) or in the depressions created when trees fall over and roots are pulled out of the soil (E.D. Brodie III 1994, personal communication). Other species of amphibians and reptiles use gaps in addition to other forest microhabitats, but at least these two species appear to use this microhabitat in ways that make them comparable to gap-specialist birds and insects. Similarly, the presence of a thick and tall canopy and its associated epiphytes has allowed many genera to develop arboreal specializations. Thus, although the majority of amphibian and reptilian species are in or on the leaf litter of the forest floor, a distinctive subset has invaded the third dimension created by the forest structure.

Seasonal patterns of climate at La Selva (Sanford et al. 1994) correlate with some aspects of herpetofaunal diversity, especially in relation to patterns of reproduction. This is particularly evident for amphibians, many of which have peaks in reproductive activity associated with the heavy rainfall at the beginning of the rainy season, or with the formation of pools of water in small streams during the dry season (Donnelly and Guyer 1994; Scott and Limerick 1983). Activity patterns, especially in frogs, are correlated with the activities of a variety of vertebrate predators, especially snakes (Greene 1988). All of these factors appear to aid in the maintenance of herpetofaunal diversity.

Despite intensive surveys at La Selva, new taxa continue to be discovered. We anticipate that new species will continue to accumulate from two sources. One source will be the pool of secretive species with geographic ranges that include La Selva but which have escaped detection. Because of their lifestyle, we expect that many of these taxa will be snakes. Only the collections of Michael J. Corn during 1969–1970 at Rio Frio, a site 20 km southeast of La Selva in Heredia Province, are available to suggest what some of these taxa might be. These samples were taken on property operated as a banana plantation by the (then) Standard Fruit Company. Of the 97 taxa collected at this site, five have never been observed at La Selva. These include one salamander (Alvarado's Salamander), one turtle (the Ornate Slider [*Trachemys ornata*]), and three snakes (Ruthven's Earthsnake [*Geophis ruthveni*], Short-tailed Littersnake [*Urotheca pachyura*], and Degenhardt's Scorpion-eater [*Stenorrhina degenhardtii*]). Two other taxa documented in the Rio Frio sample, the Four-lined Whiptail (*Ameiva quadrilineata*) and the Diamondback Racer (*Drymobius rhombifer*), have been recorded for La Selva on previous lists (Scott et al. 1983), but their status is problematic (see species accounts for the Four-lined Whiptail and the Diamondback Racer).

A second source of new taxa will be invasions into the area by species whose recent historical ranges have not included La Selva. The reinvasion of American Crocodiles (*Crocodylus acutus*) into the Rio Sarapiquí appears to be one example of this process via natural dispersal across native habitats (see species account for the American Crocodile). A second type of invasion will be taxa reaching La Selva by natural dispersal across habitats altered by human activity. The invasion by the Yellow-headed Gecko (*Gonatodes albogularis*) is a recent example of this mechanism. This species is known from disturbed areas along the Caribbean coast and eventually invaded La Selva along roads, fencerows, and housing developments. A final invasion process will be species transported by humans, intentionally or unintentionally, to the property. The introduction of the Green and Black Poison Frog (*Dendrobates auratus*) is illustrative of this process. This poison frog is found throughout much of the Atlantic lowlands of Costa Rica but for unknown reasons has never been observed between the Rios Sarapiquí and Sardinal. However, this photogenic animal recently was released at the Selva Verde Lodge at Chilamate and has since been observed at La Selva.

Conservation

Assessment of the persistence of the La Selva herpetofauna requires documentation of long-term patterns of abundance. To date, scant information is available that monitors such patterns, and it is available for only a handful of species (Braker and Greene 1994). The herpetofauna is representative of the lowland tropical wet forest fauna described by Savage (1982). As long as sites like La Selva, Barro Colorado Island (Panama), Parque Nacional Corcovado (Costa Rica), and Los Tuxtlas (Mexico) remain protected, replicate examples of this herpetofauna can be preserved. However, the critical question of how large each reserve must be to retain the current fauna is virtually unstudied. Because amphibians and reptiles are ectothermic, their populations can be dense, and their area requirements for survival are reduced relative to endotherms (e.g., Pough 1983). These factors indicate that current reserves could be large enough for all but the largest reptilian predators (such as the Spectacled Caiman *[Caiman crocodylus],* the American Crocodile, the Bushmaster *[Lachesis stenophrys],* and the Boa Constrictor *[Boa constrictor]*).

On these reserve areas, care must be taken to assess the impact of other large vertebrates. For example, if top mammalian predators like jaguars are lost and intermediate-sized predatory mammals (e.g., White-nosed Coati *[Nasua narica]*) are left unchecked, altered abundances of certain taxa of the herpetofauna might occur (Greene 1988). Additionally, for small, isolated reserves, the impact of humans killing snakes might lead to local extirpation of sensitive species like Bushmasters, which might have far-reaching impacts on forest structure (see species account for the Bushmaster).

Our ability to predict the long-term status of the herpetofauna at La Selva will come less from theoretical calculations than from careful attention to key elements of natural history. For example, Zimmerman and Bierregaard (1986) documented the importance of rare reproductive sites for some dendrobatids (some species of *Colostethus* and *Phyllobates*), and the Suriname Toad (*Pipa* sp.) in South America. Nearly all hylid frogs require a fish-free aquatic environment for successful reproduction. Two isolated open swamps located along the alluvial plains of the Rio Puerto Viejo are the primary reproductive sites of the hylid frogs at La Selva. It is likely that La Selva would be without several of these species, no matter what the reserve size, if these special sites were not present. Clearly, these resources require special management consideration. The restricted nest sites required by iguanas, crocodilians, and turtles represent another key resource for the La Selva herpetofauna. Currently, these sites are unknown and, therefore, unstudied.

Some species of reptiles have been exploited by local inhabitants for food and skins. The most heavily used species are iguanas (meat and eggs), large river turtles (meat and eggs), and crocodilians (meat and skins). La Selva serves both as a reservoir of potential recolonists in response to declines of these species in surrounding areas and as a place where natural history data of use for proper management can be gathered in a relatively pristine setting.

In planning for conservation efforts, care should be taken to identify and monitor the groups in need of special attention. At La Selva the emphasis should be on forms endemic to primary forest because these taxa have strong evolutionary ties to forest types that are imperiled by human activities. It seems likely, but is undocumented, that any reserve large enough to preserve top predators, like the Bushmaster, will maintain the other elements of the herpetofauna endemic to primary forest. La Selva has the good fortune to be able to draw upon previous herpetofaunal surveys to determine long-term trends in diversity and abundance for frogs, lizards, and some snakes (Donnelly and Guyer 1994; Guyer and Donnelly 1990; Lieberman 1986; Scott 1976).

Species that characterize areas of human disturbance or human habitation appear to be in no danger of extirpation or extinction. However, their expansion into forested areas could be used to monitor habitat degradation at La Selva. For example, Isaias Vargas, a long-time resident of the Sarapiquí region and employee of the La Selva station, has raised concerns regarding the expansion of the range within La Selva of the Striped Basilisk (*Basiliscus vittatus*), a species associated with areas disturbed by humans in other parts of Costa Rica. Again, documentation of such change is possible at La Selva because baseline data from previous inventories are available.

The long-term conservation of La Selva's herpetofauna has at least one particularly strong reason for hope. Nearly all of the species known to occupy the reserve before 1980 were resighted during a period of intense survey activity from 1982 to 1983. Thus, after a time of increased habitat loss and fragmentation around La Selva (Clark 1988), no obvious alteration of species richness was observed, a condition not shared with other nearby sites (Pounds and Crump 1994). This may indicate that the La Selva reserve and adjacent Braulio Carrillo National Park are sufficiently large to buffer the herpetofauna from the effects of nearby habitat destruction. Alternatively, insufficient time may have elapsed for population declines to occur. New surveys will be vital for distinguishing between these two interpretations. We hope that future revisions of this publication will document faunal persistence rather than decline.

AMPHIBIA

AMPHIBIANS ARE a class of land vertebrates with thin, scaleless skin and at most four fingers. The earliest amphibians appear in the fossil record in the mid-Carboniferous (approximately 360 million years before present), and about 4,600 living species are descended from these ancestors. Most present-day amphibians live in or near water and must stay moist. Typically this is accomplished by living in humid areas. However, amphibian skin is also glandular, and some of the skin glands are designed to produce a coat of lipids that retards water loss from the body. Other glands in the skin produce toxins that protect the animals from bacterial and fungal infections and from vertebrate and invertebrate predators.

Amphibians retain a fishlike life cycle in that most species lay eggs, typically in water or moist areas, which hatch into a larval stage that grows and transforms into an adult form. Eggs of modern amphibians are covered with a series of gelatinous coats. One of the functions of these protective layers is to regulate water flow to the embryo, and this feature has allowed all major groups of living amphibians to evolve some species that lay terrestrial rather than aquatic eggs; these often are attended by one of the parents. Additionally, all major orders of living amphibians contain some species that have internal fertilization, retain the developing eggs in the female uterine tract, and give birth to living offspring.

Larval amphibians are characterized by the presence of gills that protrude from external gill openings along the side of the head and/or neck (salamanders and caecilians) or that are located in special chambers (frogs and toads). For species in which larvae are aquatic, these gills function to extract oxygen from the water and to release carbon dioxide. Larvae of aquatic amphibians also typically lack eyelids and possess a laterally compressed tail designed to propel the larvae through the water. When such larvae reach an appropriate size, the gills are absorbed into the body, all openings associated with the gills become covered with tissue, the caudal fin is lost, and eyelids develop. Amphibians that have transformed but have not reached sexual maturity are referred to as juveniles; sexually mature individuals are adults. Larvae of some species that lay terrestrial eggs hatch into typical larvae, a process that usually requires careful placement of the nest by the female parent. Other species with terrestrial eggs pass the larval stage within the gelatinous layers and hatch as terrestrial juveniles, a process termed "direct development" and approaching the condition seen in many reptiles, all birds, and some mammals. Finally, amphibians that retain eggs in the uterine tract can have nutrient and/or gas exchange between the female parent and her developing offspring. Such species give birth to terrestrial juveniles, a process that approaches that of placental mammals.

Modern amphibians are entirely predaceous. Their teeth have pointed caps that are designed to pierce animal bodies. If a captured prey struggles and the tooth caps break off, they are replaced through growth of new caps. Most juvenile and adult amphibians consume invertebrate prey. However, the largest frogs and salamanders can consume small vertebrates, occasionally including birds and mammals. Because amphibians can be extremely abundant and often are easily captured, they are prized prey for other vertebrate predators. Some snake and bird species specialize on frogs; specialist predators on salamanders and caecilians are unusual.

Key to the Orders of Amphibians at La Selva

1a   Body limbless............................. Gymnophiona
1b   Body with four limbs 2
    2a   Body of adult has a tail..................... Caudata
    2b   Body of adult without a tail .... Anura

GYMNOPHIONA

Members of the order Gymnophiona are limbless, burrowing amphibians that look similar to large earthworms and are referred to as caecilians. Only about 170 species of caecilians have been described, and all of them are restricted to the tropics; they are absent from New Guinea and Australia. These animals possess several unique or unusual features that make them distinctive. Primary among these is the presence of a tentacle, a unique paired structure located near the orbits. The tentacle is small and usually requires a hand lens to see. It has an anatomical configuration that suggests that it serves to detect chemicals within the burrow and that it senses the structure of the burrow through touch. Because vision is generally of limited use to burrowing vertebrates, caecilians have degenerate eyes. These degenerate eyes probably function only to indicate whether the animal's head is in the darkened environment of the burrow or protruded into the brighter environment outside the burrow. A similar function is performed by specialized, light-sensitive organs found on the tail of caecilians. These structures are found in no other vertebrate group.

Caecilians have internal fertilization, a process accomplished with the phallodeum, a specialized copulatory structure found in males that is inserted into the cloacal opening of females during reproduction. Once fertilized, some species produce aquatic eggs, the larvae of which are free swimming. Other species retain offspring in the uterine tract of the female, where nutrients are provided and/or gas exchange takes place. Still other species lay eggs in a terrestrial nest within which the larval stage develops and from which transformed juveniles emerge (direct development).

Because of their burrowing habits, caecilians are difficult to study, and much of their life history is unknown. All adults and larvae appear to be predaceous, consuming a variety of terrestrial, subterranean, and aquatic invertebrates. Little is known regarding major predators of caecilians; however, some species are brightly colored, suggesting that skin toxins may make these animals a poor meal for predators.

In the lowland tropical wet forests of Central America, most caecilians are encountered during the wet season either by rolling logs or finding them on the ground in areas saturated with rainwater. These animals can burrow with surprising speed, so you must react quickly to capture them. Caecilians can be picked up safely with bare hands. See the description for the order Anura for tips on the short-term storage of live individuals.

Caeciliidae

With about 100 species, the family Caeciliidae is the most successful family of caecilians from the perspective of species richness. The family is widely distributed, spanning lowland wet forests and dry forest sites from Veracruz, Mexico, through Amazonian Brazil to the Rio Plata region of Argentina in the New World, and equatorial Africa, India, and the Seychelles Islands in the Old World. All members of the family produce live young. This is the only family of caecilians known from Costa Rica. Additionally, only one species occurs at La Selva, so characteristics given in the species account cover those needed to distinguish this family. One additional species is known from elsewhere in the Caribbean lowland tropical wet forests of Costa Rica.

PURPLE CAECILIAN   *Gymnopis multiplicata*
Pl. 1 
OTHER COMMON NAMES: Cecilido, Soldas,
Suelda con Suelda, Dos Cabezas, Veragua
Caecilian
SIZE: Largest individual 490 mm SVL (n = 1).
DISTRIBUTION: Guatemala to Panama.
IDENTIFYING FEATURES: The only species of caecilian known from La Selva, this animal is characterized by an elongate, limbless body, barely visible eye spots (pale white spot), and a minuscule, retractile tentacle located at the anterior edge of the eye (within the pale white spot). The body is uniform dark maroon to purple, changing immediately to the white belly coloration at the ventral quarter of body. The traditional herpetological de-

scription, that this creature is a large earthworm that bites, is an apt one, although the gape of the jaws of this animal is too small to be felt by a human.

SEXUAL DIMORPHISM: We know of no external features that distinguish the two sexes.

HABITAT: The Purple Caecilian burrows in loose soil. At La Selva, this species probably is associated with new and old alluvium. It can occasionally be found by turning over logs and old trail boards, especially after major floods resulting from heavy rains. Otherwise, its fossorial habits hide it from view.

DIET: At La Selva this animal is known to consume formicids and coleopterans (Lieberman 1986). Elsewhere it has been reported to consume a wide variety of invertebrates, dominated by oligochaetes but also including isopteran and other insect larvae and early instars (Wake 1983).

REPRODUCTION: Fertilization is internal and is effected by the eversion of the copulatory structure (phallodeum) of a male into the cloaca of a female. The female retains larvae in the oviduct; the larvae use specialized teeth to scrape the oviductal lining. Cells and secretions from the lining nourish the developing offspring. The female gives birth to from two to 10 young (Wake 1983). Nothing is known of the timing of reproduction.

REMARKS: The location of the tentacle should be carefully inspected and noted on any specimen from La Selva because this species is quite similar in general appearance to the La Loma Caecilian *(Dermophis parviceps)* (see under "Additional Species").

CAUDATA

The order Caudata comprises the living salamanders of the world and has about 520 species. Because most salamanders have four limbs and a tail, they retain a general appearance similar to that of the earliest land vertebrates. However, these features are also observed in many lizards, causing these two groups frequently to be confused with each other. No external morphological feature of adult salamanders is unique to this order. Therefore, salamanders and lizards are best differentiated by their skin: salamanders have smooth, moist, scaleless skin, and lizards have dry skin covered by epidermal scales.

Salamanders have external or internal fertilization, depending on the family. Those with external fertilization breed in a fashion that is similar to that of stream-dwelling teleost fishes. Males of such species establish a nest site within which females deposit eggs. Males follow close behind and fertilize the eggs with a cloud of sperm. Those forms having internal fertilization transfer sperm by means of a sperm packet (spermatophore) produced by the male and placed on the substrate; as a female crosses over the packet, she picks up the sperm and places it in her cloacal opening by action of the smooth muscles that line the opening. Because the spermatophore is relatively small, males must follow an elaborate set of courtship behaviors designed to induce the female to follow closely behind the male and improve the chances of sperm transfer. When the female is appropriately positioned, she signals the male to deposit the spermatophore by nudging his cloaca or pelvic girdle with her nose. As the male and female walk forward in single file, the female's body crosses the spermatophore just left by the male.

Internally fertilized females may lay eggs in an aquatic or terrestrial nest or may retain eggs in the uterine tract and give birth to live young. Most species lay eggs, and in several species the female parent attends the nest until the offspring hatch. For species that live near streams or pools of water, aquatic larvae hatch from the eggs; these larvae may spend one to several years growing to the size at which metamorphosis takes place. Several families of aquatic salamanders become sexually mature without fully transforming to the adult form, a process known as paedomorphosis. For species that live away from aquatic habitats, females lay eggs in terrestrial nests, and the larval stage takes place inside the eggs, protected by the gelatinous coating. No free-swimming larvae are produced. Instead, small fully transformed versions of the adult hatch from the eggs, a process referred to as direct development.

All adult salamanders are predaceous, consuming aquatic and terrestrial invertebrates and, among large species, small fish and other salamanders. Larvae are also predaceous, eating aquatic invertebrates and, in some cases, other salamander larvae. Although salamanders are quite common in some habitats, relatively few predators specialize on them, perhaps because they hide effectively in vegetation or organic debris. Nevertheless, a few species of snakes consume salamanders as the primary component of their diet.

Salamanders in the lowland tropical wet forests of Central America are difficult to find. Employ the following two techniques to encounter them. (1) Examine leaf surfaces of understory shrubs and epiphytic plants. This is best done at night with the aid of a flashlight, headlamp, or lantern. Typically, misting rainfall is needed before animals venture out into such exposure. (2) During daylight hours, remove leaf litter from plots of land or the bases of tree buttresses, roll logs and other forest debris, and carefully inspect mosses and the bases of leaves of bromeliads and other epiphytic flowering plants.

You can pick up salamanders safely with bare hands. Although these animals generally move slowly, they may jump or writhe vigorously when disturbed, and so they can be more difficult to capture than it might seem. The tail may become detached during such struggles. See the description for the order Anura for tips on capture and short-term storage of live individuals.

Plethodontidae

The members of the family Plethodontidae are referred to as the lungless salamanders because functional lungs are absent and are replaced by vascularized skin, especially on the lips and lining of the mouth. Gas exchange takes place across this skin rather than in lungs. This is the most species-rich salamander family, with approximately 270 living species. The family has three relatively distinctive centers of diversity, one in the Appalachian region of the United States, one in the coastal forests of the Pacific Northwest of the United States (but including two disjunct species found in Italy), and one in Mexico and Central America. The last includes species that have extended southward, invading the Amazonian basin. These Neotropical forms belong to the subfamily Bolitoglossinae, characterized by mushroom-shaped tongues and direct development, among other features. Development takes place in jelly-coated eggs laid in moist terrestrial nests. A completely metamorphosed individual emerges from the jelly coat. Three members of this family inhabit La Selva. Three additional species are known from elsewhere in the lowland tropical wet forests of Costa Rica.

Key to the Plethodontidae at La Selva

1a   Ridge across head between eyes; body short, tail about as long as rest of body; limbs well developed . . . .
. . . . *Bolitoglossa colonnea*

1b   No ridge across head between eyes; body elongate, tail two to three times length of rest of body; limbs reduced 2

    2a   Tail two times length of rest of body (use only if tail is unbroken); snout rounded; hind limbs relatively stout, slightly thicker and longer than forelimbs (fig. 2a) . . .
. . . . *Oedipina cyclocauda*

    2b   Tail three times length of rest of body (use only if tail is unbroken); snout pointed; hind limbs relatively thin, as thick and long as forelimbs (fig. 2b) . . . .
. . . . *Oedipina gracilis*

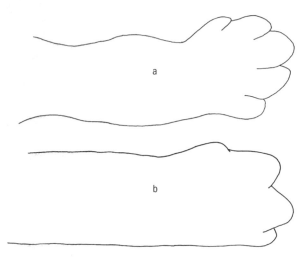

Figure 2. Conformation of hind limbs of *Oedipina* from La Selva:
(a) *O. cyclocauda;* (b) *O. gracilis.*

RIDGE-HEADED SALAMANDER   *Bolitoglossa colonnea*
Pl. 2

OTHER COMMON NAMES: Salamandra,
La Loma Salamander

SIZE: Largest individual 37 mm SVL (n = 3).

DISTRIBUTION: Caribbean slopes of Costa Rica and
central Panama; Pacific slopes of the Osa Peninsula,
Costa Rica, and adjacent Panama.

IDENTIFYING FEATURES: This genus (and species) can be distinguished from the other genus of salamander at La Selva by the presence of a relatively short trunk, short tail, and robust limbs (the Round-tailed Worm Salamander *[Oedipina cyclocauda]* and the Long-tailed Worm Salamander *[O. gracilis]* have an unmistakably long, wormlike body and wispy limbs). The body color varies from tan to black, and the color pattern may be uniform or mottled. The fingers and toes are encased in webbing so that the digits are barely visible (sometimes the feet are lighter in color than the rest of body). Also, there is an elevated ridge between the eyes.

SEXUAL DIMORPHISM: The male develops pronounced premaxillary teeth when reproductive. These teeth are used to scrape the female's skin, which allows chemicals from the male's chin gland to be delivered to the female during courtship.

HABITAT: The Ridge-headed Salamander can be found by searching moist leaves of shrubs during rainy nights, sifting through leaf litter, or examining epiphytic plants, especially the axils of bromeliad leaves.

DIET: This salamander is presumed to consume small arthropods.

REPRODUCTION: Eggs are laid in moist areas, presumably in arboreal sites. Nothing is known of the timing of egg laying or of potential parental care.

REMARKS: Two other species of this genus might occur at La Selva. These are Alvarado's Salamander *(B. alvaradoi)* and the Striated Salamander *(B. striatula)* (see under "Additional Species"). The ridge between the eyes of the Ridge-headed Salamander distinguishes this species from the other two. This characteristic should be examined carefully on any specimen of the genus *Bolitoglossa* from La Selva.

ROUND-TAILED WORM SALAMANDER and LONG-TAILED WORM SALAMANDER

Oedipina cyclocauda and *O. gracilis*

Pl. 3

OTHER COMMON NAMES: Costa Rica Worm Salamander *(O. cyclocauda)*

SIZE: Largest Round-tailed Worm Salamander 43 mm SVL (96 mm TL; n = 2); largest Long-tailed Worm Salamander 45 mm SVL (170 mm TL; n = 7).

DISTRIBUTION: Round-tailed Worm Salamander: Caribbean slopes from northern Honduras to central Panama. Long-tailed Worm Salamander: Caribbean slopes of Costa Rica to western Panama.

IDENTIFYING FEATURES: The worm salamanders are easily distinguished by their elongate, wormlike bodies and extremely long tails (two or more times the length of the body, when complete). The legs are tiny and easily overlooked. Because the two species are so similar in morphology, color, and habitat preference, we describe them together. Both species are uniform dark brown, gray, or black. For individuals with unbroken tails, the Long-tailed Worm Salamander has a much longer tail (three times the snout-to-vent length) than the Round-tailed Worm Salamander (two times the snout-to-vent length). However, tails frequently are broken and regenerated, making this characteristic of limited value in identification. The regenerated portions of such tails have a slightly lighter color than the unbroken portions. The Round-tailed Worm Salamander tends to be a more robust animal than the Long-tailed Worm Salamander in that the head is slightly wider, the snout more rounded, and the hind limbs thicker and longer. However, these differences are subtle and difficult to determine in the field unless the two species are viewed side by side.

SEXUAL DIMORPHISM: Presumably the male develops enlarged premaxillary teeth when reproductive, a characteristic of other plethodontids.

HABITAT: Both species can be found in thick, damp leaf litter or in and under rotting logs.

DIET: The diet is presumed to include small arthropods.

REPRODUCTION: Eggs presumably are laid under leaf litter or logs. Nothing is known of the timing of egg laying or of potential parental attendance at the nest.

REMARKS: The scientific names of these two species were recently corrected for specimens from La Selva (Good and Wake 1997). Older literature uses *O. uniformis* for *O. gracilis* and *O. pseudo-uniformis* for *O. cyclocauda*.

ANURA

The order Anura comprises the frogs and toads. With over 4,800 living species, this is the most species-rich group of modern amphibians and among the most successful groups of land vertebrates. Anurans are unmistakable because their bodies are adapted to a jumping mode of locomotion. In order to generate enough power for takeoff during each jump, anurans have greatly elongate hind limbs and enlarged leg muscles. Even the ankle bones of frogs are elongate, creating an extra segment to each hind leg. Adult anurans have no tail. Instead, the tail vertebrae have become modified into a bony structure, the urostyle, that solidifies the attachment of the pelvic girdle to the vertebral column. This functions to transmit the power of the hind legs to the body during a jump and creates the large bump on the back of an anuran near the junction of the legs and the body. The vertebral column of anurans is reduced to nine or fewer elements, creating a globose body shape in adults. This body plan differs strikingly from the elongate bodies of salamanders and caecilians. Finally, the shoulder girdle of anurans is loosely attached to the vertebral column, creating a shock absorber when an anuran lands after each jump.

Most anurans are vocal creatures. Males produce calls designed to establish themselves in a chorus of rival males and to attract females. Each female typically selects carefully from among calling males, allowing a selected male to grasp her from above. In most anurans, grasping occurs either in the armpit region (pectoral amplexus) or the groin region (pelvic amplexus). A few species are known to have internal fertilization. These species retain eggs in the uterine tract and give birth to live young. However, most species exhibit external fertilization; the female deposits eggs in a nest where they are covered by sperm deposited simultaneously by the male. The nest is sometimes placed in water, in a hole dug in the ground, under leaf litter, or on vegetation. In most cases the nest is left unattended. However, in several independent lineages of anurans, one or both parents stay near or on the eggs, preventing them from drying, and possibly defending them against predators and parasites.

The ancestral reproductive mode in anurans involves placement of eggs in an aquatic medium, followed by hatching and growth of the larvae, called tadpoles, and metamorphosis of the tadpoles into juveniles with the adult body form (McDiarmid and Altig 1999). Tadpoles differ from adults in body shape, most notably in having a large, muscular tail. Once they reach an appropriate size, tadpoles transform rapidly from the larval shape to the adult shape. The physiological and behavioral changes that occur are equally dramatic. For example, an individual changes from breathing through gills to breathing with lungs, and from consuming algae or small aquatic invertebrates to being predaceous on relatively large prey.

In many anuran species, eggs are placed in a terrestrial or arboreal nest and hatch into juveniles, a process called direct development. This is the most common reproductive mode of modern anurans and has evolved several times independently within many anuran families. There are numerous reproductive modes that are intermediate between the ancestral mode and direct development. These alternative modes suggest that there has been selection on anurans to avoid laying eggs in water and instead to lay them on land. Some species create nests on leaves above water, a behavior referred to as leaf breeding. In these species, tadpoles hatch from the eggs and fall into pools of water below the nests. In other forms, eggs are placed in foam nests created by the male and female parents. Eggs and tadpoles develop in these nests before entering the aquatic medium. In still others, eggs and tadpoles are transported in brood pouches or in the mouth by the male or female parent until they transform into juveniles. The great diversity of living anurans is related in part to the wide variety of reproductive modes that have evolved within this order.

All adult anurans are predaceous. Most small species consume small invertebrates, largely insects. To this diet, larger species add small vertebrates of all major groups. The largest anuran species of a particular wetland area frequently consumes small species of anurans using the same reproductive site. Tadpoles are sometimes predaceous, consuming mostly aquatic invertebrates but occasionally eating other tadpoles, or herbivorous, scraping algae or detritus. Because anurans are often abundant and easily captured, many predators specialize on consuming them. These predators frequently concentrate at anuran breeding sites where they may detect prey based on chemicals (most snakes), vision (many birds), or hearing (some bats).

In lowland tropical wet forests, anurans can be encountered at all times of the year and on virtually any substrate. However, most species are found in one of three habitats: leaf litter, stream-side vegetation, or epiphytic plants (especially bromeliads). Also, most anurans are nocturnal, and therefore, the best time to find them is at night. At this time, they typically perch on leaves, where their bodies appear bright in the light of flashlights, head-lamps, or lanterns. The eyes of larger species reflect a bright red color, a feature that allows location of animals otherwise hidden by vegetation, soil, or water. Additionally, males typically vocalize at night, allowing localization of their positions. Many litter an-urans climb on short herbaceous vegetation at night, whereas ar-boreal species may perch at any layer within the forest. Therefore, you should walk slowly, surveying all possible positions in the understory from ground level to as high as you can reach. Ani-mals perched high above your head can sometimes be induced to perch on a long pole or stick that is gently moved underneath the anuran. If such anurans do not reposition themselves on the stick, then they may jump to a lower perch, where they can be captured. In order to maximize the number of species observed, you should select routes that include trails through primary for-est, streams that can be waded, and open swamps. The last can yield spectacular results if you happen to be present when these sites first fill with water early in the wet season.

By day, litter anurans often can be found by walking trails or shuffling through areas of thick leaf litter and searching for indi-viduals that leap to avoid footsteps. Additionally, litter anurans can be found by raking leaves from the bases of tree buttresses. Anurans that occur along streamsides can be found by wading through creeks and looking through leaves backlit by sunlight. Streamside anurans often perch on such leaves, and the outlines of their bodies appear as shadows on the illuminated leaves. Fi-nally, anurans can be found during the day by examining mats of mosses, orchids, or ferns on logs or tree limbs and the leaf axils of bromeliads.

Anurans are most easily captured by hand. For individuals perched on leaves, this can be done by clapping your hands, one above and one below the anuran (including the perch material).

Anurans on the ground, rocks, or tree trunks can be pinned under a flattened hand. These animals frequently dive under leaf litter before you can attempt capture. Occasionally these can be caught by immediately pinning the leaf litter with two hands (to cover as much area of the litter as possible) and then carefully sifting the area pinned. Although all anurans have poison glands in their skin, none of the species from the lowland wet forests of Central America requires gloves for protection during capture. However, care must be taken to avoid transferring chemicals from anuran skin to sensitive areas such as your eyes, nose, or mouth. It is always best to release individuals immediately after capture at the place where they were captured. But if you want to retain individuals, then it is best to keep captured animals separated by species to avoid contact of particularly noxious species with other kinds of anurans (see the species account for the Smoky Jungle Frog *[Leptodactylus pentadactylus]*). Captured animals are best kept in plastic containers with moist vegetation or leaf litter. However, moistened cloth bags or plastic bags with moistened leaf litter (opened and closed twice daily to replenish fresh air) can be used for short periods of time.

Key to the Families of Anura at La Selva

1a Has a pair of parotoid glands behind eyes (fig. 3); these are small and difficult to distinguish from large "warts" in some species and most juveniles .... Bufonidae

1b No parotoid glands behind eyes 2

2a Body brightly colored with red and dark blue, black with metallic blue green blotches, or black with two dorsolateral, yellow or turquoise stripes .... Dendrobatidae

2b Body variously colored but never as above (some species of this couplet choice will have red on arms, legs, or groin or light, dorsolateral stripes but not in association with blue or black colors) 3

3a Thin fold of skin across back of head behind eyes (fig. 4); body short and wide with a relatively short head; venter black with large white spots and/or blotches .... Microhylidae

3b No fold of skin across back of head between eyes; body not short and wide; venter variously colored but if patterned with light and dark then ground color is light (white to olive brown) with large black spots and/or blotches 4

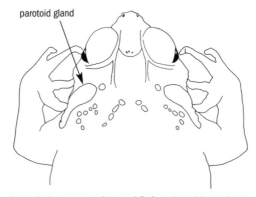

Figure 3. Dorsal view of head of *Bufo melanochlorus* showing location of parotoid glands in the family Bufonidae.

Figure 4. Dorsal view of body of *Gastrophryne pictiventris* showing location of fold of skin across back of head.

4a   Extensive webbing between hind toes, attaching half way or more along length of each hind digit (fig. 5a). 5

4b   Webbing between hind toes at most a thin wisp, attaching less than half way along length of each toe (fig. 5b) . . . . Leptodactylidae

5a   All digits with well-developed, rounded disks at tips (fig. 6a); intercalary cartilage (look carefully, many *Eleutherodactylus* have toe disks but never have intercalary cartilage, a structure that allows toe disks to bend at odd angles when held in hand). 6

5b   Toe disks absent; no intercalary cartilage . . . . Ranidae

6a   Venter transparent; internal organs or a white guanine sheet easily visible through belly skin (organs may be wrapped by a white guanine coat) . . . . . Centrolenidae

6b   Venter not transparent; internal organs not easily visible (except for bones, which may be visible) . . . . Hylidae

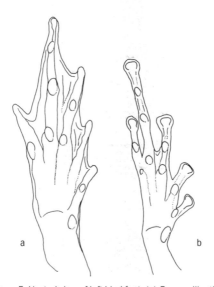

Figure 5. Ventral view of left hind foot: (a) *Rana vaillanti,* showing extensive webbing between toes; (b) *Eleutherodactylus fitzingeri,* showing limited webbing between toes.

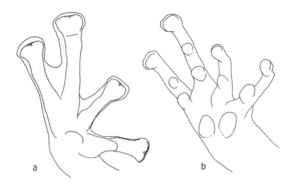

Figure 6. Ventral view of right front foot: (a) *Hyla loquax*, showing well-developed toe disks on all fingers; (b) *Eleutherodactylus fitzingeri*, showing toe disks on outer two fingers only.

Bufonidae

The Bufonidae comprises the toads, a family that contains approximately 380 species and is found on all major continents except Australia (one species introduced) and Antarctica. The unique feature used to unite all members of this family is the presence of a Bidder's organ, a remnant of the female reproductive system retained in male toads. Within the family, various species may be found in virtually any habitat type and display any reproductive mode, including live birth. In the lowland wet forests of Costa Rica, toads are easy to distinguish based on the presence of an enlarged poison-secreting gland located behind each eye (parotoid gland). Additionally, toads usually have warty skin, as well as short hind legs that are used to hop rather than leap. All species of toads in the lowland wet forests of Costa Rica migrate to aquatic sites to breed, where mated pairs deposit eggs as long strings in areas of standing water. Four members of this family occur at La Selva, and they represent the fauna known from the rest of the Caribbean lowland wet forests of Costa Rica.

Key to the Bufonidae at La Selva

1a   Skin smooth, warts not easily visible .... *Bufo haematiticus*
1b   Skin rough, warts large and easily visible 2

2a Parotoid gland huge, almost as long as head; tarsal fold (fig. 7) . . . . *Bufo marinus*

2b Parotoid gland shorter than head length; no tarsal fold 3

3a First finger (thumb) short, nearly as short as shortest finger; body green or tan with green blotches; usually without a thin, light, middorsal stripe . . . . *Bufo coniferus*

3b First finger (thumb) long, nearly as long as longest finger; body mottled with gray and dark brown; usually with a thin, light, middorsal stripe . . . . *Bufo melanochlorus*

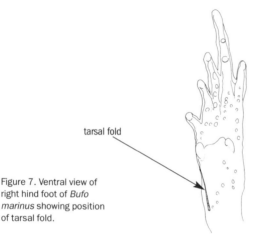

Figure 7. Ventral view of right hind foot of *Bufo marinus* showing position of tarsal fold.

tarsal fold

GREEN CLIMBING TOAD   *Bufo coniferus*
Pl. 4

OTHER COMMON NAMES: Evergreen Toad
SIZE: Largest male 60 mm SVL (n = 1); largest female 79 mm SVL (n = 1). Leenders (2001) indicates the male can be up to 72 mm SVL, and the female can be up to 94 mm SVL.
DISTRIBUTION: Caribbean slopes of Costa Rica and Panama; Pacific slopes from Costa Rica to Ecuador.
IDENTIFYING FEATURES: This toad is characterized by a light tan, brown, gray, or red dorsum, often with green blotches. The parotoid glands are small, triangular, and can be difficult to distin-

guish from body warts. The irises are gold to bronze in color with black reticulations. Each side of the body contains a row of tall, pointed tubercles aligned from behind the parotoid gland to the groin. The only similar species is the Wet Forest Toad *(B. melanochlorus)*, but the two are easily separated based on the length of the first finger (thumb); this is the shortest finger in the Green Climbing Toad and nearly the longest in the Wet Forest Toad.

SEXUAL DIMORPHISM: Our size data are limited but are consistent with the female being larger, on average, than the male, a pattern observed in other toads. Additionally, during the breeding season the male develops dark cornified patches on the dorsal surface of the thumbs, and thick, muscular forearms.

HABITAT: This species is found on the ground or in vegetation in primary and secondary forest, usually near large temporary or permanent ponds. It is primarily nocturnal but is sometimes active by day during the breeding season.

DIET: Presumably this toad eats arthropods.

REPRODUCTION: Reproduction typically occurs during the dry season. The male calls by producing a prolonged, slow trill from the edge of pools of water along streams or ponds. The call is similar to but higher in pitch than that of the Marine Toad *(B. marinus)*. Large choruses may occur day and night and sound like gaspowered engines.

REMARKS: This species is rarely encountered at La Selva and may be more common at slightly higher elevations on the Caribbean slopes of Costa Rica.

LITTER TOAD   *Bufo haematiticus*
Pl. 5

OTHER COMMON NAMES: Sapito Hojarasca, Smooth-skinned Toad, Black-belly Toad, Turando Toad

SIZE: Largest male 43.5 SVL (n = 62); largest female 54 mm SVL (n = 7). Elsewhere in Costa Rica this species reaches sizes of ca. 80 mm SVL.

DISTRIBUTION: Caribbean slopes from Honduras to northern Colombia; Pacific slopes from the Osa Peninsula, Costa Rica, to Ecuador.

IDENTIFYING FEATURES: The adult Litter Toad has a dark brown to purplish gray dorsum, often with a thin, middorsal, light (yellow)

stripe. A wide, black eye mask occurs from the tip of the snout along the sides of the body to the groin. Dorsally, this black marking is separated from the middorsal coloration by thin, yellow dorsolateral stripes (one on each side) from the nostril, over each eye, to the groin. The groin region is mottled with orange yellow, and the venter is dark gray anteriorly shading to light gray posteriorly. The hands and feet are dusty orange, and the iris is black with gold flecks. When observed in the wild, its color pattern makes this toad indistinguishable from leaf litter. No other toad at La Selva looks like this species.

SEXUAL DIMORPHISM: The adult female is presumably larger than the male. Additionally, the male develops thickened pads at the base of the thumb and appears to have thicker, more muscular forearms than the female.

HABITAT: This species is a common member of the leaf litter community and can be found in primary and secondary forests. It appears to be diurnal.

DIET: This toad eats a variety of arthropods including coleopterans, araneids, acarines, hymenopterans, and nonspider arachnids, but the diet is dominated by ants (Lieberman 1986).

REPRODUCTION: The male has a single vocal slit leading to a vocal sac, features that suggest it can vocalize. However, we have never heard this toad call, despite having observed hundreds of them. In Panama, the call of this species is described as a series of uniform, high-pitched chirps that resemble the sound of a baby chick (Ibáñez et al. 1999). Nothing is known of mate attraction. At La Selva, breeding occurs during the dry season when eggs are laid in pools along drying streams.

MARINE TOAD   *Bufo marinus*
Pls. 6, 7

OTHER COMMON NAMES: Sapo Grande, Cane Toad, Giant Toad, Agua Toad
SIZE: Largest individual 86 mm SVL (n = 13). Elsewhere this species can reach sizes of ca. 230 mm SVL (Leenders 2001).
DISTRIBUTION: Caribbean slopes from southern Texas through Central America to northern and Amazonian South America.

IDENTIFYING FEATURES: This is the largest anuran at La Selva and one of the largest in the world. As an adult, the Marine Toad is uniform light, grayish tan with huge parotoid glands that dominate the back of the head and neck. Small individuals are dark gray with considerable light gray mottling, and the parotoid gland is often difficult to distinguish. The presence of pointed tubercles on the bottoms of the fingers and toes and a fold of skin along the posteroventral portion of the tarsus (fig. 7) distinguishes the Marine Toad from all other toads at La Selva. The iris is gold with black flecks.

SEXUAL DIMORPHISM: The female can be considerably larger than the male. Also, the male develops swollen and darkened patches of skin at the base and on top of the thumbs during the breeding season.

HABITAT: The adult is principally nocturnal, although it can occasionally be observed on leaf litter by day. Often this toad is found feeding on insects attracted to lights at any of the station buildings. Additionally, it can be found in open pastures and in the leaf litter of secondary forest associated with recent alluvial soils.

DIET: This toad consumes a wide variety of invertebrates and some vertebrates.

REPRODUCTION: Adult males call during most months of the year. However, the largest choruses occur in large rivers and streams during the dry season. During the wet season some males call from small streams. The low, prolonged, rattling call carries for long distances, with chorusing commencing at dusk. Amplectant females deposit long strings of up to 25,000 eggs (Leenders 2001) in pools and backwater areas of the calling sites. The tadpoles are small, black, and often form schools along the edges of pools. At transformation, this toad has a snout-to-vent length of 8 mm.

REMARKS: Because this toad is large and its parotoid glands proportionately larger than those of other toads at La Selva, the milky, toxic parotoid secretions should be handled carefully. Psychoactive compounds are present in these secretions and may have been used by native cultures (Lee 1996); however, nausea and headaches also are known to occur in humans who have handled these animals (Allen and Neill 1956), and at least one death has been ascribed to consumption by a human (Dioscoro 1952).

WET FOREST TOAD   *Bufo melanochlorus*
Pls. 8, 9

OTHER COMMON NAMES: Dark Green Toad
SIZE: We have no measurement of this
species from La Selva. Elsewhere in Costa Rica
this toad is known to reach sizes of ca. 100 mm
SVL.

DISTRIBUTION: Caribbean slopes of Costa Rica.

IDENTIFYING FEATURES: The basic body color of the Wet Forest
Toad is brown or gray with a series of irregular, dark gray to black
blotches. The head is brown dorsally and has a distinct crest running from each nostril, behind the eye, to the back of the head.
Below each eye is a squareish, tan patch, extending to the upper
lip, that separates a mottled dark brown region extending from
the eye to the tip of the snout from a wide black patch extending
from behind the eye to the ventral edge of the parotoid gland.
This gives the appearance of a black eye mask. The iris is coppery
bronze with a black reticulum. The legs are mottled with light
and dark brown. A thin, light, middorsal stripe is present on most
individuals. The venter is dark gray to brown with light flecks anteriorly and light gray with gray spots posteriorly. A series of distinct white spots is present along the ventral edge of the lower
jaw. The juvenile is more brightly colored than the adult (Leenders 2001). A row of low, rounded, lateral warts extends from the
parotoid gland to the level of the groin. The adult is most similar
to the Green Climbing Toad *(B. coniferus)* in size. However, the
Wet Forest Toad has a first finger that is nearly as long as the third
(longest) finger, whereas the first finger is shorter than the others
in the Green Climbing Toad. The juvenile Wet Forest Toad is similar to the young Marine Toad *(B. marinus)* in color pattern. Because the parotoid glands in the juvenile of the latter species are
difficult to see, the two species must be distinguished from each
other based on differences in the shapes of the tubercles on the
bottoms of the fingers and toes (pointed in the Marine Toad;
rounded in the Wet Forest Toad) and of the posteroventral portion of the tarsus (a fold of skin in the Marine Toad, no fold of
skin in the Wet Forest Toad).

SEXUAL DIMORPHISM: The female is larger than the male. During
the breeding season the male develops a dark, thickened patch on
the dorsal surface of each thumb and has thicker, more muscular
forearms than the female.

HABITAT: This is a rare toad at La Selva and is more common at higher elevations. The only individuals that we know of from La Selva have come from along the Camino Experimental Sur, the Sendero Tres Rios, and the Sendero Holdridge trails and were observed during the dry season.

DIET: Presumably this toad eats invertebrates.

REPRODUCTION: No data are available on breeding habits from the Atlantic slopes of Costa Rica, but calling males from the Osa Peninsula produce a short trill, given during the dry season from pools along rocky streams (Savage 2002).

Hylidae

The family Hylidae contains the treefrogs. With approximately 740 species, this is a notably rich family. The family is distributed in a wide variety of habitats, from rainforests to deserts, and is found on all continents except Africa and Antarctica. Members of this family are characterized by enlarged toe disks on all digits of the hands and feet, an intercalary cartilage associated with each toe disk (see fig. 8), extensive webbing between the toes (and often the fingers), and belly skin that is at most translucent but usually opaque. In the lowland wet forests of Costa Rica, many species breed in open swamps that lack predatory fish. However, during periods of seasonal flooding, fish may invade these pools, causing the failure of entire cohorts of eggs and tadpoles (Donnelly and Guyer 1994). Other hylids breed in pools along rocky streams. Hylid frogs are nocturnal, with activity commencing at dusk, peaking shortly thereafter, and tapering off by midnight (Winter 1987). During the reproductive season, females select males based on the quality of their calls, preferring males with deeper voices, prolonged calls, and loud vocalizations (Gerhardt 1994). Additionally, females may select males based on the quality of their territory (Kluge 1981). Once mated, females place eggs in water or on leaves above water, and the male fertilizes the eggs externally. Twelve species of hylids are present at La Selva. Two additional species are known from elsewhere in the Caribbean lowland wet forests of Costa Rica.

Key to the Hylidae at La Selva

1a Sides of body boldly marked with yellow, orange, or blue vertical bars or sides of body uniform light purple; dorsum leaf green; iris red, orange, yellow, or gray 2

1b Body variously colored but without bold markings on sides; iris variously colored (usually black, tan, or coppery brown) but not as above 4

   2a Iris gray bordered by yellow; thighs and sides of body orange with dark bars . . . . *Agalychnis calcarifer*

   2b Iris red to orange; thighs and sides of body not orange with dark bars 3

3a Sides of body blue with vertical yellow bars . . . .
. . . . *Agalychnis callidryas*

3b Sides of body purple without vertical yellow bars . . . .
. . . . *Agalychnis saltator*

   4a Bones green (visible by inspecting ventral surface of arms or legs) 5

   4b Bones not green (not easily seen; inspect ventral surface of arms or legs) 6

5a Webs between fingers and toes red or orange . . . .
. . . . *Hyla rufitela*

5b Webs between fingers and toes yellow green . . . .
. . . . *Scinax elaeochroa*

   6a Posterior surface of thigh boldly marked with vertical dark and light yellowish bars . . . . *Scinax boulengeri*

   6b Posterior surface of thigh variable but not marked with bold, vertical, dark and light bars 7

7a Posterior surface of thigh red . . . . *Hyla loquax*

7b Posterior surface of thigh yellow, orange, tan, brown, or purple 8

   8a Posterior surface of thigh yellow or orange 9

   8b Posterior surface of thigh tan, brown, or purple 10

9a Body with bold, dark brown blotches bordered by bright yellow, ivory, gray, or orange . . . . *Hyla ebraccata*

9b Body with faint markings of tan and brown . . . .
. . . . *Hyla phlebodes*

   10a Ventrolateral margin of forearm with a series of short but distinct warts (fig. 8) 11

   10b Ventrolateral margin of forearm without a series of warts 12

Figure 8. Lateral view of left forearm of *Smilisca baudinii* showing position of ventrolateral warts (stippled) and the effect of the intercalary cartilage on the orientation of each toe disc.

11a A brown or black bar from below eye to margin of upper lip (may fade in captivity); anterior surface of thigh and groin yellow with bold dark bars . . . . *Smilisca baudinii*
11b No black bar below eye; anterior surface of thigh and groin purplish brown . . . . *Smilisca sordida*
   12a Webbing between fingers; posterior thigh yellowish tan . . . . *Smilisca phaeota*
   12b No webbing between fingers; posterior thigh dark brown or purple . . . . *Smilisca puma*

GRAY-EYED LEAF FROG   *Agalychnis calcarifer*

Pl. 10

OTHER COMMON NAMES: Splendid Leaf Frog
SIZE: Largest male 64 mm SVL; largest
female 78.5 mm SVL (Duellman 1970).
DISTRIBUTION: Caribbean slopes from Honduras to
Colombia; Pacific slopes from Panama to Ecuador.
IDENTIFYING FEATURES: Arguably, this large hylid is the most attractive frog at La Selva. The adult has a dark, leaf green dorsum, and the sides of the body are orange with green to purplish bars, sometimes with an indistinct reticulum. The shanks, forearms, and webbing of the hands and feet are orange. The chin is yellow, and the rest of the venter is orangish yellow. The iris is gray with a yellow border. A well-developed flap of skin, or calcar,

is found on each heel. In the juvenile, the body is olive to grayish green, with orange on the venter and orange webbing of the hands and feet (Donnelly et al. 1987). This species is not likely to be confused with any other frog at La Selva.

SEXUAL DIMORPHISM: The female is typically larger than the male, and during the breeding season, the male develops a dark brown pad at the base of each thumb.

HABITAT: This is one of the few species of La Selva amphibians or reptiles that can be considered a gap specialist (Roberts 1994a). It is found most frequently in treefall openings, where breeding takes place. The same gaps may be used in consecutive years (Marquis et al. 1986), but activity is extremely sporadic.

DIET: Presumably this frog eats arthropods.

REPRODUCTION: Males form choruses away from standing water but near fallen trees in primary forest. Such choruses have been observed in dry and wet season months (Caldwell 1994; Marquis et al. 1986). The call of each male is a faint "whuunk." Within a chorus, males may call in a consistent sequence (Marquis et al. 1986). Amplectant females swim in and then select nest sites above pools of water in depressions on the trunks of fallen trees or above pools of water that form in the holes created when trees fall and pull roots out of the soil. Nests consist of 13 to 54 eggs deposited on vegetation or bark (Caldwell 1994; Donnelly et al. 1987; Duellman 1970; Roberts 1994a). The female is presumed to absorb bladder water while swimming, and the bladder contents are then released on the clutch to hydrate it (Roberts 1994a). The eggs are turquoise when first laid, but change to yellowish brown as development takes place (Donnelly et al. 1987). Eggs hatch in five to 10 days, unless inundated by water and/or infected with fungi (Roberts 1994a). Early tadpole growth is rapid but later slows; forelimbs appear at about 50 days, and transformation occurs at 64 days (Roberts 1994a). At transformation, this frog is 23 to 27.5 mm SVL. Development within the egg mass may take as long as a month, and transformation may take as long as 169 days in laboratory culture (Donnelly et al. 1987).

RED-EYED LEAF FROG

Agalychnis callidryas

Pl. 11

OTHER COMMON NAMES: Rana Calzonudo,
Gaudy Leaf Frog, Red-eyed Treefrog
SIZE: Largest male 58 mm SVL (n = 9);
largest female 93 mm SVL (n = 1).
DISTRIBUTION: Caribbean slopes from Veracruz, Mexico, to Colombia; Pacific slopes from Nicaragua to eastern Panama.

IDENTIFYING FEATURES: This species is a boldly marked frog that is a favorite among photographers. The dorsal ground color is light leaf green, and the venter is white. Some individuals possess one to several raised yellow spots on the dorsum. The sides of the body have a series of wide, purplish blue bars separated from the dorsal and ventral coloring by bright yellow. The upper arm and thigh are purplish blue, except for a thin, green line on the anterior surface of the thigh. The webbing of the hands and feet is orange, and the iris is a bright red. The lower eyelid typically has a reticulum that is gold in color (Leenders 2001). During the day this frog flattens itself against leaves and develops light tan bands, giving it a tiger-striped appearance. Superficially, this species looks similar to the Parachuting Red-eyed Leaf Frog *(A. saltator)*. However, the color on the flanks, a uniform purple in the Parachuting Red-eyed Leaf Frog, distinguishes the two species.

SEXUAL DIMORPHISM: The female is, on average, larger than the male. During the breeding season, the male develops a dark pad at the base of each thumb.

HABITAT: At La Selva, both the Cantarana and Research swamps, as well as the pastures near the main entrance, are used by this species for breeding. This frog is arboreal and moves with a lemurlike hand-over-hand locomotion. An opposable thumb assists it in gripping thin twigs and vines. By day this species moves to perches high in the canopy trees surrounding the swamps. During periods of reproduction, it descends toward a breeding pool starting at dusk.

DIET: Presumably this frog eats arthropods.

REPRODUCTION: Breeding typically takes place throughout the rainy season. Calling males may be heard on rainy and non-rainy nights; however, most eggs are laid after intense rain showers. The male typically gives a short, single-note "chuck" call at widely spaced intervals from the vegetation surrounding open swamps. Occasionally double-note calls are given. Calling begins shortly after dusk from perches high in the vegetation, and, on nights suitable for reproduction, the male continues to call as it descends to perches around the swamp edge. The amplectant female descends to the pool and absorbs bladder water before ascending the vegetation (Pyburn 1970) and selecting a nest site on a leaf—often on the undersurface of *Panicum* or *Spathiphyllum*—where 11 to 104 turquoise eggs are laid and fertilized (Roberts 1994a). A breeding pair sometimes lays eggs before standing water is available, in anticipation of approaching storms (Donnelly and Guyer 1994). The female, along with her amplectant male, may descend back to the pond to absorb water and remoisten the eggs with bladder fluids before leaving the reproductive site (Scott 1983a). Embryos develop for about eight days before hatching as tadpoles, which then drop into the water. The length of time spent in the egg capsule may be shortened if predatory snakes such as the Northern Cat-eyed Snake *(Leptodeira septentrionalis)* disturb the clutch (Warkentin 1995). Tadpoles of this species are distinctive in that they swim in a vertical, head-up position in the middle of the water column.

REMARKS: Spurrell's Leaf Frog *(Agalychnis spurrelli)* is a species similar to the Red-eyed Leaf Frog found in the Osa Peninsula and extreme southeastern Costa Rica. Therefore, those using our book at sites near the Panama border should compare any individual that keys to the Red-eyed Leaf Frog with the species description for Spurrell's Leaf Frog (see under "Additional Species").

PARACHUTING RED-EYED LEAF FROG
Agalychnis saltator

Pls. 12, 13

OTHER COMMON NAMES: Misfit Leaf Frog
SIZE: Largest male 52 mm SVL (n = 30); largest female 69 mm SVL (n = 2).
DISTRIBUTION: Caribbean slopes from Honduras to Costa Rica.
IDENTIFYING FEATURES: This species is the least spectacularly colorful of the leaf frogs; however, it makes up for this with some unusual reproductive behaviors. The adult is a light or dark leaf green, sometimes with faint, tannish or darker green bands, forming tigerlike stripes. Some individuals have one to

several raised yellow spots on the dorsum. The sides of the body are bluish purple, as are the anterior and posterior surfaces of the thighs. The iris is orange to tomato red, the hands and feet are orange, and the venter is cream anteriorly with a yellow or orange tint posteriorly. At transformation, the juvenile is copper in color and has purplish gray sides, yellow to brown coloring on the posterior region of the thighs, and a white venter; the iris of small juveniles is yellow to light orange. This species is most similar to the Red-eyed Leaf Frog *(A. callidryas)*, but the latter species differs in the color of the flanks, which are blue with yellow bars.

SEXUAL DIMORPHISM: As with most frogs, the female is slightly larger than the male. During the breeding season, the male develops a dark or black patch on the inside of the base of the thumb.

HABITAT: The Cantarana and Research swamps are the primary reproductive sites for this species at La Selva. Presumably adults migrate to canopy trees in the primary forest around these sites during the dry season.

DIET: Presumably this frog eats arthropods.

REPRODUCTION: Unlike other leaf frogs, the male of this species produces a short, soft, high-pitched "peep" or "chirp" while at the breeding site; the call is repeated at long intervals, about 30 seconds. Calls typically are produced from perches in trees surrounding the breeding site (Strieby 1998). Breeding takes place explosively over very short time periods, typically during or immediately after intense rains, and involves large, writhing masses of frogs. Most frogs are in pairs, but a female may have up to four males attempting to amplex her (Roberts 1994b). Several females may lay their eggs in contiguous masses on the surfaces of large lianas above open swamps. The female lays a mean of about 170 eggs, and, unlike in other leaf frogs, the yolks of these eggs are gray (Roberts 1994b). The tadpoles of this species swim in a vertical, head-up position in the water column.

REMARKS: When migrating to the swamps, the Parachuting Red-eyed Leaf Frog has the unusual behavior of parachuting from canopy trees to the vegetation surrounding the reproductive site. The frog jumps from trees and lands on the leaves of understory plants (Roberts 1994b).

HARLEQUIN TREEFROG
Pls. 14, 15

Hyla ebraccata

OTHER COMMON NAMES: Hourglass Treefrog
SIZE: Largest male 28 mm SVL (n = 29);
largest female 36 mm SVL (n = 14).
DISTRIBUTION: Caribbean slopes from Veracruz,
Mexico, to Panama; Pacific slopes from Costa Rica to
Panama.

IDENTIFYING FEATURES: By night, this is a relatively drab frog with a tan dorsum mottled with darker brown. By day, however, it is boldly marked: the dorsal ground color is yellow, ivory, or orange, much of it with a brownish cast, dissected by bold chocolate brown to brick red blotches. These colors have the shiny appearance of enamel paint. A wide, dark brown stripe extends from the tip of the snout through the eyes to the level of the axilla, becoming thin posterior to the tympanum. The posterior surface of the thigh is uniform orangish tan or yellow, and the iris is tan, yellow, reddish brown, or dark brick red. The venter is lemon yellow, and the ventral surfaces of the thighs are orange. This species is most likely to be confused with the San Carlos Treefrog *(H. phlebodes)*, but the latter has a faint reticulum of gray brown on the dorsum, rather than the bold colors seen in the Harlequin Treefrog.

SEXUAL DIMORPHISM: The male can be distinguished from the female by the presence of a yellowish, baggy vocal pouch. The female is larger than the male.

HABITAT: This species is most likely to be found at its breeding sites. The male calls from vegetation, most often the stems of *Spathiphyllum* leaves, around the edge of large open swamps like the Cantarana and Research swamps. Calling typically occurs at night but may occur during daylight hours during heavy rain showers. When not reproducing, the adult may be found on vegetation in the understory surrounding a breeding site.

DIET: No studies of diet have been performed; presumably this species eats arthropods.

REPRODUCTION: Breeding occurs principally during the rainy season. At this time the male gives a prolonged "wreeek" call, often followed by a series of one to three "eek" notes given in rapid succession; this call is similar to, but deeper in pitch than, the call of the San Carlos Treefrog. Adjacent calling males synchronize their calls (Wells and Schwartz 1984) and, on nights when many males are vying for position in the chorus, may wrestle for calling sites. The female approaches a calling male with a series of short leaps, during which the male continues to call. When the female is within about 30 cm, the intensity of the male's vocalizations increases until the female orients her flanks toward the male; at this time the male ceases calling and clasps the female (pectoral amplexus; Miyamoto and Cane 1980b). The female appears to discriminate among calling males as she enters a chorus and can extricate herself from amplexus (Morris 1991). Noncalling males may take positions near calling males and attempt to intercept females (Miyamoto and Cane 1980a). The amplectant female selects a nest site on the undersurface of a leaf above water, usually within about .5 m of the water surface. The eggs number between 15 and 296 (Roberts 1994a) and have small black centers when first laid. The embryos turn light brown while developing, giving the nest the appearance of small seeds clustered on a leaf. Egg production tends to peak during intense rain storms but can occur at other times. Tadpoles are strikingly patterned with a black basal body color and gold, red, and/or white blotches. This gives each tadpole the appearance of a small poeciliid fish.

REMARKS: In addition to orienting to the calls of conspecific males, the female Harlequin Treefrog also orients to the calls of San Carlos Treefrog males (Backwell and Jennions 1993). However, the female Harlequin Treefrog may use the pulse repetition rate, the dominant frequency, and the duration of calls to distinguish males of her own species from San Carlos Treefrog males (Wollerman 1998). The two species appear to use different wetland areas to reproduce at La Selva.

SWAMP TREEFROG   *Hyla loquax*
Pls. 16–18
OTHER COMMON NAMES: Loquacious Treefrog, Mahogany Treefrog
SIZE: Largest male 45 mm SVL (n = 7).
DISTRIBUTION: Caribbean slopes from the Isthmus of Tehuantepec, Mexico, to Costa Rica.
IDENTIFYING FEATURES: The dorsal body color of this medium-sized hylid typically is uniform light lemon yellow when found at the reproductive site; it usually turns tan with brownish blotches when kept in captivity. The webbing of the hands and feet is

tomato red. The sides of the body and the venter are lemon yellow, and the iris is coppery. The key field characteristics for identifying this species are the tomato red coloring on the groin and posterior surface of the thigh and the red webbing in each armpit. The Swamp Treefrog is distinguished from the San Carlos Treefrog *(H. phlebodes)* and the Harlequin Treefrog *(H. ebraccata)* by its larger size and is distinguished from all species of the genus *Smilisca* by the red coloration on the back of the thigh.

SEXUAL DIMORPHISM: Although we have not examined average body sizes at La Selva, elsewhere this frog is unusual in displaying no size dimorphism between the sexes (Lee 1996). However, the male has baggy, yellow skin under the chin where the vocal sac expands during calling. Additionally, the male has a swollen, but unpigmented, bump at the base of each thumb.

HABITAT: At La Selva this species is found almost exclusively on grasses in the center of open swamps such as the Cantarana and Research swamps.

DIET: Presumably this species consumes arthropods.

REPRODUCTION: Breeding occurs throughout the rainy season, and calling peaks during intense rainstorms. Choruses in this species never reach the size of those of other members of the genus *Hyla*. The male calls principally from thick grass, such as *Panicum,* giving a short, deep, grating, single-note "gack." This call is reminiscent of a short, gagging cough and is repeated at short intervals, about every five seconds. The amplectant female deposits a mean of 690 eggs in water (Roberts 1994a).

SAN CARLOS TREEFROG   *Hyla phlebodes*
Pl. 19

OTHER COMMON NAMES: Veined Treefrog
SIZE: Largest male 23 mm SVL (n = 1);
largest female 24.5 mm SVL (n = 1).
DISTRIBUTION: Caribbean slopes from southern
Nicaragua to Panama; Pacific slopes from eastern
Panama to Colombia.

IDENTIFYING FEATURES: This frog has a tan ground color with a reticulum of darker gray brown. A thin, dark brown stripe extends from the tip of the snout to the eye in most individuals. The posterior surface of the thigh is uniform yellow, the iris is bronze, and the venter is white. At night, this small, drab treefrog looks very similar to the Harlequin Treefrog *(H. ebraccata)*. However, its coloring remains unchanged when viewed by day, unlike the Harlequin Treefrog, which, by day, is boldly marked with deep enamel yellow and dark brown blotches.

SEXUAL DIMORPHISM: The female is probably larger than the male, and a calling male probably has baggy skin under the chin during the reproductive season.

HABITAT: The male San Carlos Treefrog calls from vegetation in the Cantarana and Research swamps, as well as from pools in pastures near the La Selva station's family housing. In our experience, this species is more common in open, disturbed areas than other hylids of similar size (the Harlequin Treefrog and the Olive Snouted Treefrog *[Scinax elaeochroa]*).

DIET: Presumably this frog eats arthropods.

REPRODUCTION: The call of the male typically consists of two slow notes ("wreek") followed by a series of five to eight shorter notes ("eek"), which are higher in pitch and not as loud as the notes of the Harlequin Treefrog. Reproduction occurs sporadically throughout the wet season, often associated with periods of particularly heavy rain. A mean of about 400 eggs is laid in water (Roberts 1994a), typically in small clusters floating in vegetation (Duellman 1970).

REMARKS: The other common name that we list for this species is from Leenders (2001). However, this common name is listed by Liner (1994) and Lee (2000) for *Phrynohyas venulosa*, a large hylid from drier sites. Therefore, we do not follow Leenders (2001) in the use of the Veined Treefrog as the common name for *H. phlebodes*.

SCARLET-WEBBED TREEFROG — *Hyla rufitela*

Pls. 20–22

OTHER COMMON NAMES: Canal Zone Treefrog

SIZE: Largest male 43 mm SVL (n = 2); largest female 54 mm SVL (n = 3).

DISTRIBUTION: Caribbean slopes from Nicaragua to Panama.

IDENTIFYING FEATURES: This frog rivals the leaf frogs (genus *Agalychnis*) in beauty. In the adult, the dorsum is either lime or bluish. White (or blue) spots and/or small, black flecks may be added to the dorsal ground color. The groin, anterior and posterior thigh, and finger and toe webbing are deep red to orange. The iris is gray in the center, shading to yellow peripherally. In the juvenile, the dorsum is lime green anteriorly and more yellow posteriorly. A red stripe extends along each side of the body from the tip of the snout, over the eye and tympanum, to the vent. A wider, yellow stripe is below the red one. Another yellow stripe runs along the anal fold. The limbs are lime green dorsally, and the posterior surface of the thigh is green. The webbing on the hands and feet is orange. The chin is bluish white, and the rest of the venter is white or yellow. This species has green bones, a feature most easily seen through the undersurfaces of the limbs.

SEXUAL DIMORPHISM: The female is larger than the male. Additionally, the reproductive male develops a spikelike bump on the medial surface of the thumb, midway from the base to the tip.

HABITAT: The Scarlet-webbed Treefrog is sometimes found on the leaves of understory plants in primary forest, but it is most consistently seen and heard in a swamp downslope from the trailmarker at CES 500 where large calling choruses aggregate. This site is characterized by very shallow water surrounded by swamp forest vegetation. This species is an occasional visitor to the Research swamp.

DIET: Presumably this species eats arthropods.

REPRODUCTION: Like all La Selva hylids, the Scarlet-webbed Treefrog selects fishfree habitats in which to reproduce. Calling males can be heard sporadically during the wet season. The call is a high-pitched "cluck" or "quack" (weak in volume relative to most other members of the family) that sounds like a soft duck call. The note is repeated in a series that lasts about five seconds, with long pauses of about five minutes between each series. Calls are given from secluded positions in vegetation close to the water surface, usually from the cyclanth cola de gallo *(Asplundia uncinata)*. Eggs are deposited in water as a surface film (Duellman 1970).

REMARKS: The species displays great color variation. No explanation has been given for this variation. However, ontogenetic or sex-related patterns are suspected.

BOULENGER'S SNOUTED TREEFROG   *Scinax boulengeri*

Pls. 23, 24

OTHER COMMON NAMES: Ranita de Boulenger,
Boulenger's Treefrog

SIZE: Largest male 41 mm SVL (n = 2);
largest female 48 mm SVL (n = 5).

DISTRIBUTION: Caribbean slopes of Nicaragua, Costa
Rica, and Panama; Pacific slopes from Costa Rica to
Ecuador.

IDENTIFYING FEATURES: For a hylid, this species has an unusual
body shape and is, therefore, easily distinguished from other
treefrogs. The body is strongly flattened dorsoventrally, and the
snout is pointed and appears upturned because the nostrils are
on a protuberant projection. The body is covered by numerous
low, warty projections. These features give the animal a "dead
leaf" appearance. Accentuating this appearance is a dorsum that
is mottled with several shades of brown, brick red, and/or lime
green. Unique among La Selva's frogs, the posterior surface of
each thigh consists of alternating vertical bands of dark brown
and light yellow or reddish yellow. The groin region is yellow
green mottled with black, and the posterior venter is light gray or
cream with gray spots. The anterior venter is pale yellow, shading
to cream under the chin. The chin has several small, gray spots,
and the iris is copper in color.

SEXUAL DIMORPHISM: Presumably the female is larger than the
male.

HABITAT: We have observed this species only at two breeding sites,
the Cantarana and Research swamps. The male calls from grasses
in open swamps (Strieby 1998), usually with the head facing down
and cocked away from the substrate. This posture is thought to
be characteristic of the genus (Duellman and Wiens 1992).

DIET: Presumably this frog eats arthropods.

REPRODUCTION: The call of the male is a guttural, prolonged
"wraak," deeper in pitch than in the Olive Snouted Treefrog *(S.
elaeochroa)* and given over widely spaced time intervals of 10 seconds to several minutes. Calls are given from secluded locations
on vegetation. Breeding takes place in the wet season, and the
breeding season is prolonged. Calling males are more numerous
on rainy nights but can be heard on dry nights as well. The amplectant female lays a mean of about 740 eggs in water (Roberts
1994a).

REMARKS: We follow Pombal and Gordo (1991) for the generic designation of this frog; previous literature refers to it as a member of the *Hyla rubra* group or the genus *Ololygon.*

OLIVE SNOUTED TREEFROG   *Scinax elaeochroa*
Pls. 25, 26

OTHER COMMON NAMES: Sipurio Snouted Treefrog

SIZE: Largest male 32 mm SVL (n = 4); largest female 38 mm SVL (n = 3).

DISTRIBUTION: Caribbean slopes from Nicaragua to western Panama; Pacific slopes from the Pantarenas, Costa Rica, to Panama.

IDENTIFYING FEATURES: The adult Olive Snouted Treefrog is small in size and, when not breeding, has a dorsal color pattern that consists of diffuse brown stripes on a light gray to olive ground color; the venter is light gray. An individual with these colors is typically captured away from a breeding site. When calling at a breeding site, the male is bright yellow. Noncalling males and females from near a breeding site are uniform olive on the dorsum, shading to yellowish green on the sides, and the venter is light yellow to cream. In all individuals the iris is coppery orange above, shading to gray below. This species is most similar to the San Carlos Treefrog *(Hyla phlebodes)* but differs from that species in dorsal coloration: the Olive Snouted Treefrog has a reticulum whereas the San Carlos Treefrog has stripes. Also, the Olive Snouted Treefrog has blue green bones, a feature easily seen through the ventral skin; the bones of the San Carlos Treefrog are white.

SEXUAL DIMORPHISM: The female is only slightly larger than the male.

HABITAT: This is the "bathroom frog" of La Selva: the species is often found in toilet bowls or tanks or on bathroom walls, which the frog traverses with great speed in a strange "wiggling" gait reminiscent of gecko locomotion. Dispersed males occupy tree holes or trail poles throughout much of La Selva.

DIET: This frog is known to eat arthropods, such as adult and larval coleopterans (Lieberman 1986).

REPRODUCTION: Breeding takes place early in the wet season when large open swamps such as the Cantarana and Research swamps first fill (Donnelly and Guyer 1994). At this time, huge choruses of males congregate during and immediately after massive rain showers. The male turns a brilliant yellow and remains at the reproductive site day and night. The call of the male consists of one to several "waaack" notes, higher in pitch than and lacking the grating quality of those of Boulenger's Snouted Treefrog *(S. boulengeri)*. The call is typically given from vegetation around the edges of open swamps, such as from *Spathiphyllum* at the Research and Cantarana swamps (Strieby 1998). With several hundreds to thousands of males calling simultaneously, a constant din can be heard from these areas for periods of 24 to 48 hours. Males leap from site to site on the vegetation in the swamps, apparently grappling for calling perches. Smaller choruses occur for several days afterward. During this period, calling by males may be led by a dominant male that begins a chorus and is then joined by several other males (Duellman 1967b). The amplectant female deposits a mean of about 980 eggs in the water (Roberts 1994a). After this brief bout of reproductive activity, the frogs disperse from the reproductive sites. Away from the breeding sites males often call singly from trail poles or small holes in trees. These calls typically consist of a single note.

REMARKS: We use the generic designation of Pombal and Gordo (1991) for this frog; previous literature refers to it as a member of *Hyla* or *Ololygon*.

COMMON MEXICAN TREEFROG
Pl. 27

Smilisca baudinii

OTHER COMMON NAMES: Baudin's Treefrog, Mexican Smilisca, Mexican Treefrog, Van Vliet's Frog, Tooter

SIZE: Largest male 67 mm SVL (n = 1; Roberts 1994a); largest female 88 mm SVL (n = 2).

DISTRIBUTION: Caribbean slopes from southern Texas to Costa Rica; Pacific slopes from Sonora, Mexico, to central Costa Rica.

IDENTIFYING FEATURES: The Common Mexican Treefrog has a variable ground color of pale green to brown, usually with dark brown dorsal markings. A brown or black stripe runs from the tip of the snout posteriorly to the insertion of the arm; this stripe is usually visible on the frog immediately after capture, but the portion anterior to each eye may fade if the animal is kept in captivity, leaving only a dark stripe behind each eye. The dark lip bar is often bordered posteriorly by a wide, squareish, light (white to greenish yellow) spot. The groin has a reticulum of bold black and white or light yellow markings. The anterior thigh surfaces are yellow to tan with dark bars, and the posterior thigh surfaces are brown with small, cream spots. The dorsal half of the iris is golden, and the ventral half is gray. This species is most likely to be confused with the Masked Treefrog *(S. phaeota)*, but the Common Mexican Treefrog has a suborbital bar (which may fade in captivity) and a light mark on the upper lip. Additionally, the Masked Treefrog has a weblike reticulum of thin gray markings in the groin, which is unlike the bold light and dark markings in the Common Mexican Treefrog.

SEXUAL DIMORPHISM: The female is larger than the male. Additionally, the calling male has paired vocal pouches associated with a pair of dark circular markings on the chin.

DIET: No diet studies have been performed at La Selva, but this frog presumably eats arthropods; grasshoppers are consumed by this species in other parts of its range (Noble 1918).

HABITAT: This large, wide-ranging species is relatively rare at La Selva. We have observed it at the Cantarana and Research swamps during or immediately after intense rainstorms. The frog often inhabits houses, where it appears to return to the same perch night after night and consumes cockroaches (O. Vargas 1998, personal communication).

REPRODUCTION: This species is an explosive breeder; chorusing occurs on only a few nights during the first strong rains of the wet season. The male gives a deep, explosive "bonk" call that is repeated two to six times in rapid succession. The clutch produced by the female is large, containing up to 2,000 eggs, and is laid as a surface film on water (Duellman 1970).

MASKED TREEFROG
Pl. 28

Smilisca phaeota

OTHER COMMON NAMES: Central American Smilisca, Tarraco Treefrog, New Granada Crossbanded Treefrog

SIZE: Largest individual 54 mm SVL (n = 1). Elsewhere in Costa Rica the male reaches sizes of ca. 65 mm SVL, and the female reaches sizes of ca. 78 mm SVL (Leenders

2001).

DISTRIBUTION: Caribbean slopes from Honduras to northern Colombia; Pacific slopes from Costa Rica to Ecuador.

IDENTIFYING FEATURES: This is a large hylid that can be distinguished from other *Smilisca* species by its size, light silver to white upper lip, and network of fine, dark markings along the flanks. Additionally, the loreal region (between the eye and the nostril) is usually green, but this feature is shared by some Common Mexican Treefrog *(S. baudinii)* individuals. A narrow, black stripe extends from the tip of the snout to the eye. Past the eye it widens to form a thick, black stripe that encompasses the tympanum and extends to the axilla, where it is connected to dark, netlike flank markings. The dark mask contrasts with the tan and green dorsal ground color. The posterior surface of the thigh is uniform yellowish tan. The iris is bronze in color. In size and color pattern, this species is most similar to the Common Mexican Treefrog, but the Masked Treefrog's unbarred lip (rather than a dark bar followed by light mark) and weblike flank markings (rather than bold light and dark marks) distinguish the two species.

SEXUAL DIMORPHISM: The female is larger than the male.

HABITAT: This species is common in areas disturbed by humans. At La Selva, some males call from storm drains around the laboratory buildings.

DIET: Presumably this species eats arthropods.

REPRODUCTION: We have not observed reproduction in this species, but elsewhere it breeds early in the wet season in open swamps. Eggs are laid in great numbers (more than 1,000 per clutch) as a surface film (Duellman 1970). The call of the male is a deep, guttural "wonk" that is given from small pools such as those found in roadside ditches and storm drains.

REMARKS: A species of "flying" treefrog, the Highland Fringe-limbed Treefrog *(Hyla miliaria),* is also known from the Caribbean slopes of Costa Rica. This species has been collected at midelevation sites on the slopes of Volcan Barba and at scattered localities elsewhere in the country. The Highland Fringe-limbed Treefrog may be present at La Selva, and, if so, it may key to the Masked Treefrog using our key. However, the species description of the Masked Treefrog does not match the distinctive characteristics of the Highland Fringe-limbed Treefrog (see under "Additional Species").

TAWNY TREEFROG   *Smilisca puma*
Pl. 29

OTHER COMMON NAMES: Tawny Smilisca,
Nicaragua Crossbanded Treefrog
SIZE: Largest individual 36 mm SVL (n = 4).
DISTRIBUTION: Caribbean slopes of southern
Nicaragua and Costa Rica.

IDENTIFYING FEATURES: The Tawny Treefrog is the smallest
and least colorful *Smilisca* species at La Selva. The dorsum is a
yellowish tan, mottled with slightly darker brown; some individuals may have a)(-shaped mark on the dorsum. A dark, interorbital bar or V-shaped mark is typically present. The Tawny Treefrog has a thin, light stripe along the edge of the upper lip, a thin,
white stripe along each forearm, and a thin, light anal stripe. The
posterior surface of each thigh is brown. The dorsal half of each
iris is tan, and the ventral half is gray. This species is most similar
in coloration to the Drab Treefrog *(S. sordida),* from which it
differs by being smaller and lighter in color, and by lacking webbing between the fingers.

SEXUAL DIMORPHISM: Presumably the female is larger than the
male.

HABITAT: We have observed this species too infrequently to under-
stand its habitat requirements at La Selva. However, most individuals that we have seen were collected in bromeliads near the
main entrance gate to the La Selva property. Elsewhere, this
species is commonly found in areas disturbed by humans (Duellman 1970).

DIET: Presumably this species eats arthropods.

REPRODUCTION: The call is described by Duellman (1970) as being
a "squawk...followed by a series of one or more rattling secondary notes." While at the breeding site, males call from the bases of
clumps of vegetation (Duellman 1970). This species breeds during the early wet season. Based upon Duellman's (1970) natural-
history description, we expect this frog to breed in open, grassy
pastures that collect water for short periods of time. We have not
observed it breeding at the Research or Cantarana swamps.

Plate 1. Purple Caecilian.

Plate 2. Ridge-headed Salamander.

Plate 3. Round-tailed Worm Salamander.

Plate 4. Green Climbing Toad (male). Photo taken in Hitoy Cerere, Costa Rica.

Plate 5. Litter Toad (female).

Plate 6. Marine Toad (female).

Plate 7. Marine Toad (juvenile).

Plate 8. Wet Forest Toad.

Plate 9. Wet Forest Toad.

Plate 10. Gray-eyed Leaf Frog (female).

Plate 11. Red-eyed Leaf Frog (male).

Plate 12. Parachuting Red-eyed Leaf Frog (male).

Plate 13. Parachuting Red-eyed Leaf Frog (juvenile).

Plate 14. Harlequin Treefrog (male and female).

Plate 15. Harlequin Treefrog (female).

Plate 16. Swamp Treefrog (female).

Plate 17. Swamp Treefrog (male).

Plate 18. Swamp Treefrog.

Plate 19. San Carlos Treefrog (male).

Plate 20. Scarlet-webbed Treefrog.

Plate 21. Scarlet-webbed Treefrog.

Plate 22. Scarlet-webbed Treefrog (juvenile).

Plate 23. Boulenger's Snouted Treefrog.

Plate 24. Boulenger's Snouted Treefrog.

Plate 25. Olive Snouted Treefrog (male).

Plate 26. Olive Snouted Treefrog (male).

Plate 27. Common Mexican Treefrog.

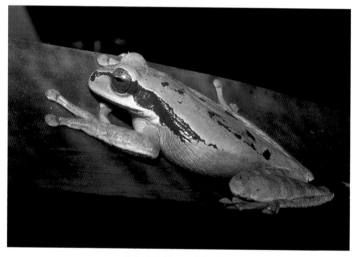

Plate 28. Masked Treefrog.

Plate 29.
Tawny Treefrog.

Plate 30.
Drab Treefrog.

Plate 31.
Emerald Glassfrog.
Photo taken in
Monteverde,
Costa Rica.

Plate 32.
Emerald Glassfrog.

Plate 33. Yellow-flecked Glassfrog.

Plate 34. Granular Glassfrog.

Plate 35. Spined Glassfrog.

Plate 36. Fleischmann's Glassfrog (juvenile).

Plate 37.
Fleischmann's
Glassfrog.
Photo taken
in Monteverde,
Costa Rica.

Plate 38.
Powdered Glassfrog.

Plate 39.
Reticulated
Glassfrog (male).

Plate 40. Coral-
spotted Rainfrog.

Plate 41. Bransford's Litterfrog (male and female).

Plate 42. Leaf-breeding Rainfrog (female).

Plate 43. Leaf-breeding Rainfrog (male).

Plate 44. Clay-colored Rainfrog (female).

Plate 45. Slim-fingered Rainfrog (female).

Plate 46. Slim-fingered Rainfrog.

Plate 47. Golden-groined Rainfrog (male).

Plate 48. Golden-groined Rainfrog (male).

Plate 49. Common Tink Frog (male).

Plate 50. Fitzinger's Rainfrog (female).

Plate 51. Fitzinger's Rainfrog (female).

Plate 52. Broad-headed Rainfrog (female).

Plate 53. Northern Masked Rainfrog (female).

Plate 54. Northern Masked Rainfrog (male).

Plate 55. Noble's Rainfrog (male).

Plate 56. Noble's Rainfrog (female).

Plate 57. Lowland Rainfrog.

Plate 58. Pygmy Rainfrog.

Plate 59. Talamancan Rainfrog (female).

Plate 60. Talamancan Rainfrog.

Plate 61.
Fringe-toed
Foamfrog (male).

Plate 62.
Fringe-toed
Foamfrog (male).

Plate 63. Smoky
Jungle Frog.

Plate 64. Smoky
Jungle Frog.

Plate 65. Smoky Jungle Frog (male).

Plate 66. Reticulated Sheepfrog (female).

Plate 67. Reticulated Sheepfrog (female).

Plate 68. Green and Black Poison Frog. Photo taken in Hitoy Cerere, Costa Rica.

Plate 69. Strawberry Poison Frog.

Plate 70. Striped Dart-poison Frog.

Plate 71. Striped Dart-poison Frog.

Plate 72. Vaillant's Frog (female).

Plate 73. Vaillant's Frog (juvenile).

Plate 74. Brilliant Forest Frog.

Plate 75. Brilliant Forest Frog.

Plate 76. Brilliant Forest Frog.

DRAB TREEFROG   *Smilisca sordida*
Pl. 30

OTHER COMMON NAMES: Veragua Crossbanded Treefrog
SIZE: Largest male 40 mm SVL (n = 7);
largest female 50 mm SVL (n = 1). Elsewhere,
the male reaches sizes of 45 mm, and the female
reaches sizes of 73 mm SVL (Leenders 2001).

DISTRIBUTION: Caribbean slopes from Honduras to Panama;
Pacific slopes from Pantarenas, Costa Rica, to Panama.

IDENTIFYING FEATURES: This is a medium-sized frog characterized
by a uniform gray brown to tan dorsum with a weakly developed,
black eye mask. A dark, interorbital bar is present in most individuals. The posterior thigh is uniform dark purplish brown with
cream, tan, pale blue, or green flecks; the groin is also dark purplish brown but has several white spots. The belly is uniform
white, changing abruptly to the darker dorsal color. The iris is
bronze with black reticulations (Duellman 1970). This species is
most likely to be confused with the other three species of *Smilisca*
at La Selva. However, the Drab Treefrog lacks (1) the green loreal
markings and network of fine, dark flank markings found in the
Masked Treefrog *(S. phaeota)*, (2) the distinctive, dark eye mask
and bold mottling in the groin found in the Common Mexican
Treefrog *(S. baudinii)*, and (3) the thin, light lip marking of the
Tawny Treefrog *(S. puma)*. Additionally, the Drab Treefrog can be
distinguished from the Tawny Treefrog by the presence of webbing between the bases of the fingers.

SEXUAL DIMORPHISM: The female is larger than the male. During
the breeding season, the male has a well-developed, black, cornified nuptial patch at the base of the thumb.

HABITAT: This frog is found in riparian zones along major rivers
and creeks. Males call from shallow backwater areas of the Rio
Puerto Viejo at the base of the steps at the River Station during
the dry season.

DIET: Presumably this frog eats arthropods.

REPRODUCTION: The male calls during the dry season, when water
levels in the river are reduced. The call is a high-pitched, pulsed,
single- or double-note "wrink" with a rising inflection at the end.
The pulses make the call sound like someone rapidly rubbing
their thumb across a short-tined comb. The call can be given
alone but is often emitted in extremely rapid succession by sev-

eral males for five to 10 seconds, followed by long periods of silence. When calling, the male selects a site in the water at the edge of a pool. The amplectant female lays her eggs in a shallow pool along the river margin. Eggs are deposited in great numbers, approximately 1,000 per clutch, as a surface film (Duellman 1970).

Centrolenidae

This family contains the glassfrogs, so named because their belly skin is transparent, allowing you to view the internal organs. In most cases, some or all of the internal organs are wrapped in a guanine coat such that the actual organ tissues are not visible, but their outlines are. In other species a white sheet of guanine associated with the peritoneal lining of the venter obscures some or all internal organs. Glassfrogs are close relatives of the treefrog family Hylidae (Ford and Cannatella 1993) and share with treefrogs enlarged disks on all digits, intercalary cartilage (see fig. 8), and extensive toe webbing. This family contains at least 120 species that are distributed from southern Mexico, through Central America, along the eastern slopes of the Andes to Bolivia, along the northern coast of South America, and along the northeastern part of Argentina. In the lowland wet forests of Costa Rica, all centrolenid species lay eggs on leaves above small streams or on rocks in streams. The male may have a projection of bone from the shoulder, called a humeral hook, or from the thumb, called a prepollical spine, that is used in wrestling matches with other males to gain favorable sites from which to attract females. These sites typically are where females lay eggs, and one or both parents may attend the nest. Six species in three genera are present at La Selva. One additional species is known from elsewhere in the Caribbean lowland wet forests of Costa Rica.

Key to the Centrolenidae at La Selva

1a Dorsum green with white or light yellow spots or punctations. 2
1b Dorsum green but without white or light yellow markings 5
  2a Bones green (visible by inspecting ventral surface of arms or legs) 3
  2b Bones white (visible by inspecting ventral surface of arms or legs) 4

3a Digestive organ tissue visible through ventral body wall, i.e., not coated with guanine; dorsum with numerous small (less than 1 mm) light spots or punctations. . . . .
. . . . *Cochranella albomaculata*

3b Digestive organ tissue white, coated with guanine; dorsum with large (more than 1 mm) light spots . . . .
. . . . *Hyalinobatrachium pulveratum*

  4a Heart white, i.e., covered by guanine coat; dorsum with numerous small, light spots that are green for more than half of the dorsum . . . .
. . . . *Hyalinobatrachium fleischmanni*

  4b Heart red, not covered by guanine coat; dorsum with large, light spots that are green for half or less of the dorsum . . . . *Hyalinobatrachium valerioi*

5a Half, or less than half, of internal organs obscured by white guanine sheet; dorsum granular; dorsal coloration bluish green . . . . *Cochranella granulosa*

5b More than half of internal organs obscured by guanine sheet; dorsum smooth; dorsal coloration lime to emerald green 6

  6a Posterior one-third of internal organs easily visible through ventral body wall, anterior two-thirds obscured by white guanine sheet; dorsum uniform green; male without humeral hooks (spikelike bony projections from the base of the humerus) but with prepollical spines (spikelike projections from the base of the first finger) . . . . *Cochranella spinosa*

  6b All internal organs hidden by a white guanine sheath; dorsum uniform emerald green, usually with black punctations; male with humeral hooks but lacking prepollical spines . . . . *Centrolenella prosoblepon*

EMERALD GLASSFROG *Centrolenella prosoblepon*

Pls. 31, 32

OTHER COMMON NAMES: Ranita de Vidrio, Nicaragua Giant Glassfrog

SIZE: Largest male 26.5 mm SVL (n = 2); largest female 27 mm SVL (n = 2). Leenders (2001) indicates 33 mm as the largest SVL for this species.

DISTRIBUTION: Caribbean slopes from Honduras to Colombia; Pacific slopes from Honduras to Ecuador.

IDENTIFYING FEATURES: This species is a small frog characterized by a dark, emerald green dorsum, usually with small, dark punctations but sometimes uniform green. The bones of this species are green, a feature that is visible through the ventral skin. A white sheet of guanine is visible through the venter, obscuring all of the organs, and the iris is ivory gray with a black reticulum. All other species of glassfrogs at La Selva are spotless or have light (blue, white, or yellow) spots or punctuations. In general color and size, this frog is most similar to the Spined Glassfrog *(Cochranella spinosa)*, a species that differs from the Emerald Glassfrog in that it lacks punctations, possesses darker green bones, and has a guanine sheet that leaves the posterior portion of the digestive system visible.

SEXUAL DIMORPHISM: The male has a distinctive humeral hook, a spikelike bony projection from the proximal end of the humerus. The female also has a spur on the upper arm, but it is much smaller than that of the male. These spurs can be used to grip opponents during encounters.

HABITAT: This species is nocturnal and is found on the leaves of trees and shrubs overhanging swift-flowing creeks.

DIET: This frog is known to eat coleopterans and orthopterans at La Selva (Lieberman 1986).

REPRODUCTION: Breeding takes place during the rainy season, when the male calls from the undersurface of a leaf that overhangs a creek. The call is a high-pitched, usually three-note whistle or chirp ("chee-chee-chee"). The amplectant female typically lays her eggs on the upper surface of a leaf in the territory of a calling male. Neither parent attends the nest. Tadpoles emerge from eggs at about 10 days of age; these tadpoles then drop into the stream where larval development is completed. The body of the tadpole is elongate, and the mouthparts are adapted for adhering to rocks at the bottom of streams.

YELLOW-FLECKED GLASSFROG
Pl. 33

Cochranella
albomaculata

OTHER COMMON NAMES: Ranita de Vidrio,
White-spotted Cochran Frog
SIZE: Largest individual 21 mm SVL (n = 1).
Elsewhere in Costa Rica this species reaches
sizes of ca. 31 mm SVL.
DISTRIBUTION: Caribbean slopes from Honduras to
Costa Rica; Pacific slopes from Costa Rica to Colombia.
IDENTIFYING FEATURES: The dorsum of this species is smooth and
uniform dark green, with numerous light silver to light yellow
punctations; these punctations are small anteriorly, becoming
noticeably larger posteriorly. The iris is silvery tan to light yellow
with a black reticulum. A white peritoneal sheath is present, and
the pericardium is covered by a guanine sheath. The rest of the
digestive system is visible through the ventral body wall. The
bones are green, a feature visible through the ventral surface. This
species is most similar to Fleischmann's Glassfrog *(Hyalinobatrachium fleischmanni)* and the Powdered Glassfrog *(H. pulveratum).* Fleischmann's Glassfrog is lighter green, has spots rather
than small, light punctations, and possesses white bones. The
Powdered Glassfrog has larger spots (more than 1 mm) and exposed digestive organs.
SEXUAL DIMORPHISM: We know of no external features that distinguish the two sexes.
HABITAT: This species is often found far from flowing water; we
have observed it in old cacao *(Theobroma cacao)* plantations
along the Rio Puerto Viejo. Presumably this species breeds along
the major rivers and streams, creating nests on leaves above
water, like other centrolenids. We have observed this frog during
daylight hours, usually jumping from understory vegetation that
we jostled, but we presume it to be largely nocturnal.
DIET: Presumably this frog feeds on small arthropods.
REPRODUCTION: Nothing is known of reproduction in this species;
the tadpole has yet to be described.
REMARKS: We follow the generic designation of Ruiz-Carranza
and Lynch (1991); older literature places this species in the genus
Centrolenella.

GRANULAR GLASSFROG *Cochranella granulosa*

Pl. 34

OTHER COMMON NAMES: Ranita de Vidrio, Grainy Cochran Frog

SIZE: We have no measurement of this species from La Selva. Elsewhere in Costa Rica this species is known to reach sizes of ca. 31 mm SVL.

DISTRIBUTION: Caribbean slopes of Honduras to Panama; Pacific slopes from Pantarenas, Costa Rica, to Panama.

IDENTIFYING FEATURES: This species differs from the other glass-frogs at La Selva in having granular skin (small bumps that require a hand lens to see); all other species have smooth skin. The dorsal ground color has a bluish cast because each bump of the granular skin has a bluish white punctation, visible with a hand lens. This color pattern makes this species readily distinguishable from all other glassfrog species at La Selva. The anterior mass of the digestive organs is not easily visible through the ventral body wall because this region is covered with a white guanine sheet. Only the posterior tip of the liver, the posterior loop of the stomach, and the large intestine are easy to see through the skin. The iris is light golden yellow with black reticulations, and the bones are green.

SEXUAL DIMORPHISM: The male develops a pad on each thumb during the breeding season, but these require a hand lens to see (Ibáñez et al. 1999).

HABITAT: This frog has been heard at the El Salto bridge along the SOR trail and has been found on leaves in primary swamp forest.

DIET: Presumably this frog feeds on small arthropods.

REPRODUCTION: The male calls from the vegetation above slow-moving streams. The call consists of a single, pulsed primary note ("creek") that may be followed by one to five secondary notes (Ibáñez 1993). Several males typically call simultaneously, each attempting to overlap the calls of the others, followed by long periods of silence (Ibáñez 1993).

REMARKS: We follow the generic designation of Ruiz-Carranza and Lynch (1991); older literature places this species in the genus *Centrolenella*.

SPINED GLASSFROG
Cochranella spinosa

Pl. 35

OTHER COMMON NAMES: Ranita de Vidrio,
Spiny Cochran Frog

SIZE: Largest individual 27 mm SVL (n = 2).

DISTRIBUTION: Caribbean slopes of Costa Rica
and Panama; Pacific slopes from Costa Rica to
Ecuador.

IDENTIFYING FEATURES: This is the smallest of La Selva's glassfrogs. It is uniform light green in dorsal coloration and does not have dark or light spots. The upper and lower lips are white or light yellow, forming a thin, light line. The major limb bones are dark green and are visible through the skin of the venter. A white sheet of guanine obscures the digestive system from the pectoral girdle to the posterior end of the stomach. The rest of the digestive system is visible through the ventral body wall. The iris is ivory gray with black reticulations. This species is most similar to the Emerald Glassfrog *(Centrolenella prosoblepon)* in body coloration, but the Emerald Glassfrog usually has small, black punctuations (sometimes lacking), lighter green bones, and a complete guanine sheet.

SEXUAL DIMORPHISM: The male possesses a small bony dagger at the base of the first finger, called a prepollical spine. The spines are used in territorial wrestling matches. In the female the prepollical region is swollen but lacks a spine.

HABITAT: We have seen this frog at night on the vegetation along the trails surrounding the Cantarana swamp. Presumably it nests on vegetation along rivers and creeks.

DIET: The diet is presumed to consist of small arthropods.

REPRODUCTION: Presumably this species lays eggs on the vegetation above streams.

REMARKS: We follow the generic designation of Ruiz-Carranza and Lynch (1991); older literature places this species in the genus *Centrolenella*.

FLEISCHMANN'S GLASSFROG
Pls. 36, 37

Hyalinobatrachium fleischmanni

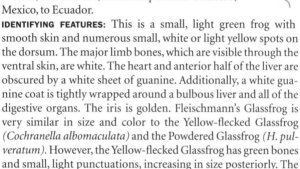

OTHER COMMON NAMES: Ranita de Vidrio, Northern Glassfrog

SIZE: Largest individual 26.5 mm SVL (n = 5).

DISTRIBUTION: Caribbean slopes from Veracruz, Mexico, to Suriname; Pacific slopes from Guerrero, Mexico, to Ecuador.

IDENTIFYING FEATURES: This is a small, light green frog with smooth skin and numerous small, white or light yellow spots on the dorsum. The major limb bones, which are visible through the ventral skin, are white. The heart and anterior half of the liver are obscured by a white sheet of guanine. Additionally, a white guanine coat is tightly wrapped around a bulbous liver and all of the digestive organs. The iris is golden. Fleischmann's Glassfrog is very similar in size and color to the Yellow-flecked Glassfrog *(Cochranella albomaculata)* and the Powdered Glassfrog *(H. pulveratum)*. However, the Yellow-flecked Glassfrog has green bones and small, light punctuations, increasing in size posteriorly. The Powdered Glassfrog has a silver iris and a fleshy forearm ridge. Both the Yellow-flecked Glassfrog and the Powdered Glassfrog differ from Fleischmann's Glassfrog in having exposed digestive organs, not ones wrapped in guanine.

SEXUAL DIMORPHISM: The female is slightly larger than the male.

HABITAT: This frog is nocturnal and is found on riparian vegetation along swift-moving streams. It can be heard calling from trees along the Rio Puerto Viejo and on the bridge above the Quebrada Sura on the trail to the River Station.

DIET: Presumably this frog consumes small arthropods.

REPRODUCTION: The male gives a high-pitched, short, single-note, whistlelike or chirplike call from the vegetation over a small, swift stream. Eggs, about 25 to 50, are laid on leaf surfaces above these streams and are attended by the male parent (Hayes 1991). The male sits on the egg mass and occasionally voids bladder water on it, thus preventing desiccation. However, if nest attendance is too frequent, fungus and fungal flies invade the egg mass, killing the embryos (Hayes 1991). The breeding season is prolonged; calling males are heard from early March to late November.

REMARKS: We follow the generic designation of Ruiz-Carranza and Lynch (1991); older literature places this species in the genus *Centrolenella*.

POWDERED GLASSFROG   *Hyalinobatrachium pulveratum*
Pl. 38

OTHER COMMON NAMES: Ranita de Vidrio, Chiriqui Glassfrog

SIZE: We have no measurements of this species from La Selva. Elsewhere in Costa Rica this frog reaches sizes of ca. 29 mm SVL.

DISTRIBUTION: Caribbean slopes from Honduras to Colombia; Pacific slopes from southeastern Costa Rica to Colombia.

IDENTIFYING FEATURES: This glassfrog is characterized by a lime green dorsum with numerous small, white spots. The arms and legs have similar coloration, but the spots are larger. The iris is silver gray, and the bones are pale green. The forearms have a fleshy flap along their margins. The venter is clear, with all viscera visible; the heart, liver, and digestive system are white because they are wrapped tightly in a guanine coat. The Powdered Glassfrog is most similar in color pattern to the Yellow-flecked Glassfrog (*Cochranella albomaculata*) and Fleischmann's Glassfrog (*H. fleischmanni*). The Yellow-flecked Glassfrog has most of its digestive organs hidden behind a guanine sheet. Fleischmann's Glassfrog has white bones and lacks the fleshy margin to the forearm of the Powdered Glassfrog. Finally, both the Yellow-flecked Glassfrog and Fleischmann's Glassfrog have yellow or golden irises.

SEXUAL DIMORPHISM: We know of no external features that distinguish the two sexes.

HABITAT: Presumably this frog is found in the vegetation along swift-moving streams and is nocturnal.

DIET: This frog is presumed to consume small arthropods.

REPRODUCTION: The male calls by producing a harsh, high-pitched "tick" that is given in a slow series (Ibáñez et al. 1999). We presume that this species breeds along swift streams and places eggs on leaves above the water. No description of parental care is available, and the tadpole of this species is undescribed.

REMARKS: We follow the generic designation of Ruiz-Carranza and Lynch (1991); older literature places this species in the genus

Centrolenella. A similar species to this one, the Ghost Glassfrog *(C. ilex),* has been recorded from Rara Avis, a private reserve located in the protected zone upslope from La Selva. Therefore, the Ghost Glassfrog might occur at La Selva. If so, it is likely to key to the Powdered Glassfrog in our key (see under "Additional Species").

RETICULATED GLASSFROG *Hyalinobatrachium valerioi*
Pl. 39

OTHER COMMON NAMES: Ranita de Vidrio, La Palma Glassfrog

SIZE: We have no measurements of this species from La Selva. Elsewhere in Costa Rica this frog reaches sizes of ca. 30 mm SVL.

DISTRIBUTION: Caribbean slopes from Costa Rica to northern Colombia; Pacific slopes from the Osa Peninsula, Costa Rica, to Ecuador.

IDENTIFYING FEATURES: This species of glassfrog is characterized by large, white to yellow spots on a lime green background. The heart is visible through the ventral skin, but the rest of the internal organs are wrapped in a white guanine coat. All major limb bones are white, and the iris is gold. No other glassfrog at La Selva looks like this species.

SEXUAL DIMORPHISM: We know of no external features that distinguish the two sexes. However, individuals attending clutches are males (McDiarmid 1978).

HABITAT: This frog frequents the vegetation along small, free-flowing streams. Activity occurs principally at night.

DIET: This frog is presumed to consume small arthropods.

REPRODUCTION: Eggs are laid on the undersurfaces of leaves in riparian settings. A single male attends one to several nests.

REMARKS: The color pattern is thought to mimic that of the egg mass being attended. We follow the generic designation of Ruiz-Carranza and Lynch (1991); older literature places this species in the genus *Centrolenella.*

Leptodactylidae

This diverse family contains approximately 900 species of frogs of a variety of shapes and sizes. The family is found in extreme southern Texas and Arizona, throughout most of Mexico and all of Central America, and extends through Amazonian South America. No single diagnostic field characteristic serves to delimit the family. Therefore, the best way to recognize these frogs is to eliminate all other possibilities. A variety of reproductive modes are expressed by leptodactylids, including laying aquatic eggs, creating foam nests in which eggs are deposited, and having direct development in eggs laid in a terrestrial or arboreal nest. Most species are found in moist places, but some are adapted to extreme aridity. In the lowland wet forests of Costa Rica, two genera occur, one that creates foam nests *(Leptodactylus)* and one with direct development *(Eleutherodactylus)*. With 16 species, this is the most species-rich family of amphibians at La Selva. Three additional species are known from elsewhere in the Caribbean lowland wet forests of Costa Rica.

Key to the Leptodactylidae at La Selva

1a   Expanded disks present on all fingers or on outer two fingers (tips of outer two fingers noticeably larger than those of inner two fingers [fig. 6b]; this is often difficult to see in small individuals) 2

1b   No expanded disks on fingers 12

    2a   A thin, membranous web present between base of toes (fig. 5b; visible by spreading toes apart; this is a small, unimpressive structure in most species). 3

    2b   No toe webs 7

3a   Posterior surface of thigh brown with small, light, yellow to tan spots (not always well formed); usually ventral surface of chin suffused with a gray reticulum except for a midventral light area (forming a light stripe) . . . .
. . . . *Eleutherodactylus fitzingeri*

3b   Posterior surface of thigh uniform brown, light purple, or red, without yellow spots; no indistinct midventral white stripe on chin 4

    4a   Upper lip white or light yellow from tip of snout to level of tympanum (often dissected by gray along an-

terior portion of lip), this in marked contrast to color
of rest of head . . . . *Eleutherodactylus talamancae*

4b   Upper lip mottled to barred with light and dark, but
not in marked contrast to rest of head. 5

5a   Posterior thigh orange or red . . . . . *Eleutherodactylus noblei*

5b   Posterior thigh brown. 6

6a   Black eye mask. . . . . *Eleutherodactylus mimus*

6b   No black eye mask. . . . . *Eleutherodactylus crassidigitus*

7a   Groin and anterior thigh surface with a golden yellow
blotch and/or oblique black bars; snout rounded in dorsal
view. . . . . *Eleutherodactylus cruentus*

7b   Not as above 8

8a   Groin and anterior surface of thigh with pink to coral
red spots or blotches bordered by dark . . . .
. . . . *Eleutherodactylus altae*

8b   Groin and anterior surface of thigh without pink to
coral spots or blotches. 9

9a   Iris pink or yellow; flaplike projection from heel (calcar)
. . . . *Eleutherodactylus caryophyllaceus*

9b   Iris not pink or yellow; no flaplike projection from heel
. . . . 10

10a   Triangular blotch between eyes (usually distinct when
first captured, may become faint in captivity); no red
on groin, thigh, or calf . . . . *Eleutherodactylus diastema*

10b   No triangular blotch between eyes; red on groin, thigh,
and/or calf 11

11a   Enlarged supraocular tubercles (fig. 9); no)(-shaped ridge
on back. . . . . *Eleutherodactylus ridens*

11b   No enlarged supraocular tubercles;)(-shaped ridge on back
. . . . *Eleutherodactylus cerasinus*

12a   A thin, membranous web present between base of toes
(fig. 5b; visible by spreading toes apart; this is an unimpressive structure) . . . . . *Eleutherodactylus ranoides*

12b   No webbing between toes. 13

13a   An X- or)(-shaped ridge on back; head unusually broad
. . . . *Eleutherodactylus megacephalus*

13b   No X- or)(-shaped ridge on back; head normal in size. . . . .
. . . . 14

14a   Red wash in groin (faint in some individuals); bottoms of hands with numerous, light-colored, pointed
tubercles. . . . . *Eleutherodactylus bransfordii*

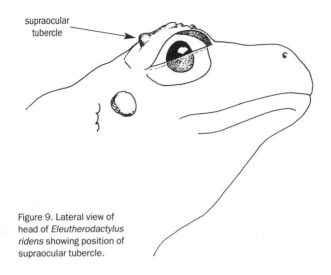

supraocular
tubercle

Figure 9. Lateral view of
head of *Eleutherodactylus
ridens* showing position of
supraocular tubercle.

14b Orange, yellow, or brown wash in groin; bottoms of
hands with dark, rounded tubercles 15
15a Lips barred boldly with light and dark; dorsum of adult has
a reticulum of dark brown and lighter reddish brown . . . . .
. . . . *Leptodactylus pentadactylus*
15b Lips not barred boldly with light and dark, at most marked
with three light spots; dorsum of adult more or less uniform
brown or gray. . . . . *Leptodactylus melanonotus*

CORAL-SPOTTED RAINFROG   *Eleutherodactylus altae*
PI. 40

OTHER COMMON NAMES: Mountain
Robber Frog

SIZE: Largest male 22 mm SVL (n = 3);
largest female 28.5 mm SVL (n = 3).

DISTRIBUTION: Caribbean slopes of Costa Rica.

IDENTIFYING FEATURES: The body of the Coral-spotted Rainfrog is short and wide with relatively short hind limbs. The color
pattern is distinctive; no other frog has a pattern of bold pink to
coral patches in the groin and on the anterior surface of the thigh.

These patches are outlined in black, delimiting them from the dark gray dorsal and light gray ventral ground colors. The Golden-groined Rainfrog *(E. cruentus)* is the only other species that has such bold markings in the groin, but these are golden rather than pink or coral.

SEXUAL DIMORPHISM: The adult male appears to be smaller than the female.

HABITAT: This frog is rarely encountered at La Selva. It perches on shrubs at night in primary forest, using moderately developed disks on the outer two fingers for climbing. It may be more common in steep ravines at the middle elevations of Braulio Carrillo National Park.

DIET: Presumably this species eats small arthropods.

REPRODUCTION: The breeding habits of this frog have not been studied. Presumably reproduction is similar to that of the Leaf-breeding Rainfrog *(E. caryophyllaceus)* because both are arboreal and of similar size.

BRANSFORD'S LITTERFROG
Pl. 41

Eleutherodactylus
bransfordii

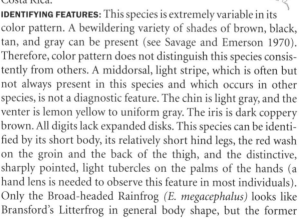

OTHER COMMON NAMES: Bransford's Robber Frog

SIZE: Largest male 23 mm SVL (n = 212); largest female 27.5 mm SVL (n = 341).

DISTRIBUTION: Caribbean slopes of Nicaragua and Costa Rica.

IDENTIFYING FEATURES: This species is extremely variable in its color pattern. A bewildering variety of shades of brown, black, tan, and gray can be present (see Savage and Emerson 1970). Therefore, color pattern does not distinguish this species consistently from others. A middorsal, light stripe, which is often but not always present in this species and which occurs in other species, is not a diagnostic feature. The chin is light gray, and the venter is lemon yellow to uniform gray. The iris is dark coppery brown. All digits lack expanded disks. This species can be identified by its short body, its relatively short hind legs, the red wash on the groin and the back of the thigh, and the distinctive, sharply pointed, light tubercles on the palms of the hands (a hand lens is needed to observe this feature in most individuals). Only the Broad-headed Rainfrog *(E. megacephalus)* looks like Bransford's Litterfrog in general body shape, but the former differs from the latter in possessing white belly spots on a dark background and a noticeably larger head. Two other species, the Clay-colored Rainfrog *(E. cerasinus)* and the Pygmy Rainfrog *(E. ridens),* have red coloration in the groin or on the back of the thigh, but both of these species have expanded toe disks, a feature lacking in Bransford's Litterfrog.

SEXUAL DIMORPHISM: The female is larger than the male. Additionally, the tympanum of the male is relatively larger (about the size of the eye) than that of the female (smaller than the size of the eye). Finally, the adult male develops a pad at the base of each thumb that is not present in the female.

HABITAT: This probably is the most abundant and widely distributed frog at La Selva, being found throughout primary and secondary forest habitats (Heinen 1992; Lieberman 1986). During the dry season this species tends to accumulate in low-lying areas near streams. This species is often captured during daylight hours. However, measurements of activity patterns suggest that small individuals are diurnal and adults are crepuscular or nocturnal (Winter 1987).

DIET: The diet is a varied collection of arthropods, including orthopterans, adult and larval dipterans, hemipterans, homopterans, dermopterans, nonspider arachnids, isopods, and centipedes; but coleopterans, araneids, acarines, and formicids are consumed in the greatest numbers (Lieberman 1986; Limerick 1976).

REPRODUCTION: Though difficult to hear, the male does give a faint, short, high-pitched call that sounds like the "cheep" of a chick. This species reaches peak abundance in the leaf litter during the dry season, perhaps indicating peak reproduction during this time period (Donnelly 1999).

LEAF-BREEDING RAINFROG
Pls. 42, 43

Eleutherodactylus caryophyllaceus

OTHER COMMON NAMES: La Loma Robber Frog
SIZE: Largest male 22.5 mm SVL (n = 9); largest female 32 mm SVL (n = 3).
DISTRIBUTION: Caribbean slopes from Costa Rica to Colombia; Pacific slopes from extreme southwestern Costa Rica to Colombia.
IDENTIFYING FEATURES: This species has dramatic color variation,

ranging from tan to dark green. Some individuals have stripes, and others have chevron-shaped markings on the dorsum. Often a dark, triangular, interorbital mark is present. The iris is either pink or yellow, and a flaplike heel tubercle, or calcar, is present. This species has moderately well-developed toe disks on the outer two fingers, a feature that allows it to climb. This frog tends to have a sleek appearance because of its long, pointed snout and elongate hind limbs. The Leaf-breeding Rainfrog looks most similar to the Pygmy Rainfrog *(E. ridens)*, but the latter species has a coppery brown iris, lacks a calcar, and possesses distinctive supraocular tubercles (a bump above each eye; fig. 9).

SEXUAL DIMORPHISM: Presumably the male is smaller than the female.

HABITAT: This frog is an arboreal denizen of primary forest. It is most likely to be found on the leaves of trees and shrubs near streams on rainy nights. The species is uncommon at La Selva but is abundant at about 1,000 m in the Zona Protectora.

DIET: At La Selva, this frog is known to consume orthopterans, dipterans, lepidopteran larvae, and isopods (Lieberman 1986).

REPRODUCTION: Eggs are deposited on leaves, and the female remains with the nest (Myers 1969b; Townsend 1996). The eggs are unusual in that the outer jelly coat becomes white and may be calcareous. Leenders (2001) suggests that reproduction occurs principally at night and that the mating call of the male is a soft "pew."

REMARKS: The dorsal color pattern and variable iris color may be related to the sex of the individual. This variation needs investigation.

CLAY-COLORED RAINFROG *Eleutherodactylus cerasinus*
Pl. 44

OTHER COMMON NAMES: Limon Robber Frog
SIZE: Largest male 22 mm SVL (n = 3);
largest female 35 mm SVL (n = 15).
DISTRIBUTION: Caribbean slopes from Nicaragua to Panama.

IDENTIFYING FEATURES: The dorsum of this species is mottled with various shades of light brown, sometimes with a reddish cast. The animal sometimes has a wide, middorsal, light stripe from the tip of the snout to the cloaca. The sides of the face have indistinct light and dark lip bars. The groin, anterior thigh, and anterior shank are red, whereas the posterior thigh is uniform dark brown. The iris is two-toned, with the upper half being golden yellow and the lower half coppery brown. The hind limbs are relatively long, and the head narrows at the tip of the snout. The outer two finger disks are noticeably wider than those of the inner two fingers. The body shape, the golden to coppery iris, and the red wash in the groin are good field characteristics. Two species have similar features: the Golden-groined Rainfrog *(E. cruentus)* has a similar body form but differs from the Clay-colored Rainfrog in having golden yellow (rather than red) in the groin and a coppery (rather than golden) iris. The Pygmy Rainfrog *(E. ridens)* has red in the groin and on the back of the thigh but differs from the Clay-colored Rainfrog in being smaller and having an enlarged supraocular tubercle on each eyelid.

SEXUAL DIMORPHISM: The male is presumed to be smaller than the female.

HABITAT: This frog can be found day or night at La Selva. It is relatively common in the Arboretum, along the banks of the Quebrada Sura. Adults perch on the ground by day and on leaves or shrubs near the ground at night; enlarged toe disks on the outer two digits of the hands aid in accessing these nocturnal perches. The frog escapes by making two to four long leaps, ending with a dive under the leaf litter.

DIET: This frog is known to consume coleopterans, dipterans, araneids, and acarines at La Selva (Lieberman 1986).

REPRODUCTION: We have not heard the male call at La Selva. In Panama, the male produces a single, short, high-pitched note ("tick;" Ibáñez et al. 1999). Presumably this frog is a ground nester.

SLIM-FINGERED RAINFROG
Eleutherodactylus crassidigitus

Pls. 45, 46

OTHER COMMON NAMES: Isla Bonita Robber Frog

SIZE: Largest male 28 mm SVL (n = 14); largest female 45 mm SVL (n = 9).

DISTRIBUTION: Caribbean slopes from Costa Rica to Colombia; Pacific slopes from Pantarenas to the Osa Peninsula of Costa Rica.

IDENTIFYING FEATURES: Like all members of the Fitzinger's Rainfrog *(E. fitzingeri)* group, the Slim-fingered Rainfrog has elongate hind limbs, a relatively pointed snout, and greatly expanded disks on the outer two fingers, and it uses three to four leaps in random directions to escape from large predators. The dorsum is light tan, and some individuals have a bold, light, middorsal stripe. The back of the thigh is brown with no spots, and the upper lip is mottled. The iris is yellow with thin brown streaks and a brown horizontal stripe. Two species at La Selva can be confused with the Slim-fingered Rainfrog. Fitzinger's Rainfrog looks virtually identical except that it has yellow spots on the back of the thigh, and the webbing of the hind foot is less extensive. The Talaman-can Rainfrog *(E. talamancae)* differs from the Slim-Fingered Rainfrog in having a light upper lip and purple to red coloring on the back of the thighs.

SEXUAL DIMORPHISM: The male is much smaller than the female. Additionally, the tympanum of the male is larger (about the size of the eye) than that of the female (smaller than the eye). Finally, the adult male develops pads at the bases of the thumbs.

HABITAT: This frog is a resident of primary forest, where it can be observed at night perched low—usually less than 1 m above the ground—on vegetation. Enlarged disks on the outer two fingers are an adaptation to this arboreal existence. By day this species is found under the leaf litter in primary forest. Measurements of activity patterns suggest that this species is most active at dawn and dusk (Winter 1987).

DIET: This animal eats orthopterans, dipterans, isopods, and cen-tipedes at La Selva (Lieberman 1986).

REPRODUCTION: The male gives a high-pitched, single-note "waah" at widely spaced intervals. Taylor (1952) reported an adult (sex not identified) under a rock with 26 eggs.

GOLDEN-GROINED RAINFROG
Pls. 47, 48

Eleutherodactylus
cruentus

OTHER COMMON NAMES: Chiriqui Robber Frog
SIZE: Largest male 28 mm SVL (n = 8); largest female 31 mm SVL (n = 3).
DISTRIBUTION: Caribbean slopes of Costa Rica and Panama; Pacific slopes from Costa Rica to Colombia.
IDENTIFYING FEATURES: This frog is characterized by enlarged toe

disks on the outer two fingers (relative to the inner two), yellow mottled with black in the groin, and a tubercle on top of each eyelid. The iris is bronze, sometimes reddish, with black reticulations. This combination of characteristics is unique among the rainfrogs at La Selva. The dorsum is often dark brown to black. The frog often has a thin, light, middorsal stripe from the tip of the snout to the cloaca. The coloring of the anterior and posterior surfaces of the thigh consists of a series of black and dark gray bars. The lips are also barred with light and dark gray. On the venter, this frog is gray with black mottling. The most similar other species are the Clay-colored Rainfrog *(E. cerasinus)* and the Pygmy Rainfrog *(E. ridens),* both of which have red on the groin and/or back of the thigh or calf, a feature lacking in the Golden-groined Rainfrog.

SEXUAL DIMORPHISM: Unlike most frogs, the female does not appear to be larger than the male.

HABITAT: The Golden-groined Rainfrog perches at night high on vegetation, usually more than 1 m above the ground; greatly enlarged disks on the toes and outer two fingers are used for climbing. This frog can be found in primary and secondary forests. It appears to prefer dense and undisturbed cloud forest or wet forest habitats.

DIET: This frog consumes hymenopterans and hemipterans at La Selva (Lieberman 1986).

REPRODUCTION: We have not observed the male to produce a call. Eggs in this species are laid in crevices on tree trunks. The parents do not attend the eggs (Myers 1969b; Townsend 1996).

COMMON TINK FROG *Eleutherodactylus diastema*
Pl. 49

OTHER COMMON NAMES: Martillito, Common Dink Frog, Caretta Robber Frog

SIZE: Largest male 23 mm SVL (n = 35); largest female 25.5 mm SVL (n = 14).

DISTRIBUTION: Caribbean slopes from Nicaragua to Colombia; Pacific slopes from central Costa Rica to Ecuador.

IDENTIFYING FEATURES: This small frog is characterized by the presence of equivalently sized disks on all digits that are slightly pointed at the anterior end. At night, the adult appears uniform light gray. By day the dorsum is brown with wide, middorsal, dark brown markings and with a dark brown, triangular marking starting between the eyes and pointing posteriorly. The venter is uniform white or mottled with light and dark gray; the male has a yellow chin, and the female has a gray chin. The iris is light brown to brick red. The most similar other species at La Selva is the Pygmy Rainfrog (*E. ridens*), which differs from the Common Tink Frog in having red in the groin and on the back of the thigh and having only two fingers with expanded disks, rather than all four.

SEXUAL DIMORPHISM: The male can be distinguished from the female by the presence of a yellow vocal sac. The adult female is slightly larger than the male.

HABITAT: This abundant frog is arboreal, living on understory plants and epiphytic vegetation, especially bromeliads. This animal is difficult to locate because it frequently calls from the undersurfaces of leaves, which makes its voice sound as though it is coming from a different location. We have heard this frog throughout La Selva, wherever there are trees with epiphytes.

DIET: This species consumes a variety of arthropods, including coleopterans, orthopterans, non-ant hymenopterans, dipterans, isopods, araneids, nonspider arachnids, and acarines, but the diet tends to be dominated by formicids (Lieberman 1986).

REPRODUCTION: This frog gets its common name from the high-pitched, metallic "tink" call that the male gives during or immediately after heavy rains, day or night. For such a common frog, surprisingly little is known about its reproduction. Up to 10 eggs are laid in bromeliads (Leenders 2001), but these do not appear to be attended by either parent.

FITZINGER'S RAINFROG　　*Eleutherodactylus fitzingeri*
Pls. 50, 51

OTHER COMMON NAMES: Fitzinger's Robber Frog

SIZE: Largest male 30 mm SVL (n = 66); largest female 48 mm SVL (n = 65). Leenders (2001) reports males up to 35 mm and females up to 52 mm.

DISTRIBUTION: Caribbean slopes from Honduras to northern Colombia; Pacific slopes from Costa Rica to Colombia.

IDENTIFYING FEATURES: This species has long hind legs and an elongate snout. The dorsum has an indistinct reticulum of brown and gray. Many individuals have a dark, interorbital bar and a W-shaped, dark mark behind the eyes. Others may have a bold, white or light yellow, middorsal stripe from the tip of the snout to the cloaca; this striped morph looks like, but is not, a separate species. The sides of the face and lips are barred with light and dark gray. The posterior surface of the thigh is dark gray brown with a light yellowish reticulum and/or spots (difficult to see in smaller individuals). The venter is white anteriorly, shading to yellow posteriorly. The iris is copper brown to golden yellow. Typically, the chin has a wide, white stripe bordered laterally by gray. This species can be distinguished from two similar species, the Slim-fingered Rainfrog *(E. crassidigitus)* and the Talamancan Rainfrog *(E. talamancae),* that have uniform color on the back of the thigh and lack a light, midventral stripe on the chin.

SEXUAL DIMORPHISM: The male is much smaller than the female. Additionally, the tympanum of the male is larger (about the size of the eye) than that of the female (smaller than the eye). Finally, the adult male develops a thickened pad at the base of each thumb.

HABITAT: This species is common throughout La Selva but especially in areas with recently deposited alluvial soils; it is especially common (or easy to see) in the Arboretum. At night, the adult can be found perched low on understory vegetation; by day this animal is found under leaf litter. The adult escapes by a series of leaps (usually three), each in a different direction, ending with a dive under the leaf litter. Measurement of activity patterns suggests that activity peaks at dawn and dusk (Winter 1987).

DIET: Presumably this frog eats arthropods.

REPRODUCTION: The call of the male consists of a series of five to seven "clack" notes given in rapid succession; these notes sound like two stones being struck together. This species nests on the ground, and the female attends the eggs (Dunn 1931; Lynch and Myers 1983).

BROAD-HEADED RAINFROG
Pl. 52

Eleutherodactylus
megacephalus

SIZE: Largest male 39 mm SVL (n = 28); largest female 71 mm SVL (n = 6).

DISTRIBUTION: Caribbean slopes from Honduras to western Panama.

IDENTIFYING FEATURES: This species has a distinctive short, toadlike body and an unusually broad head. Typically, the dorsum is light tan to gray brown. The iris is dark brown. An X- or)(-shaped ridge is etched in black on the anterior portion of the dorsum. The venter coloring consists of white spots on a dark gray to olive brown reticulum, and the iris is black with gold flecks. The digits lack toe disks. Most individuals encountered are small juveniles, but the adult can be the size of a large ranid. The only remotely similar species is the ubiquitous Bransford's Litterfrog *(E. bransfordii),* which, like the Broad-headed Rainfrog, has relatively short hind legs and dorsal ridging. However, the light spots on the belly of the Broad-headed Rainfrog distinguish the two.

SEXUAL DIMORPHISM: Two distinct size groups are evident at La Selva. Those that are 20 to 28 mm SVL include males. A group of much larger individuals is presumed to be adult females.

HABITAT: The Broad-headed Rainfrog is found by day or night on the leaf litter in old second growth and primary forest. The absence of toe disks suggests that this species does not climb on vegetation. Measurements of activity patterns suggest that small individuals are diurnal and large ones are nocturnal (Winter 1987).

DIET: At La Selva, this frog eats large arthropods (formicids and other hymenopterans, adult and larval coleopterans, orthopterans, araneids, homopterans, dermopterans, dipteran larvae, isopods) and small vertebrates (Lieberman 1986).

REPRODUCTION: Eggs are laid on the leaf litter. No vocalizations are known from this species. Juvenile abundance increases during the dry season, suggesting a late wet season bout of reproduction (Watling and Donnelly 2002).

REMARKS: At sites near the Panama border, individuals that key to the Broad-headed Rainfrog in our key should be compared with the description for the Warty Rainfrog *(E. bufoniformis)* (see under "Additional Species"). Older literature will refer to this species as *E. biporcatus.*

NORTHERN MASKED RAINFROG

Eleutherodactylus mimus

Pls. 53, 54

OTHER COMMON NAMES: Northern Mimicking Rainfrog, Tilaran Robber Frog

SIZE: Largest male 38 mm SVL (n = 52); largest female 59 mm SVL (n = 30).

DISTRIBUTION: Caribbean slopes from Honduras to Costa Rica.

IDENTIFYING FEATURES: This species is characterized by the presence of a wide, black eye mask from the tip of the snout, through the eye, above the axilla, to midway between the axilla and groin. A glandular ridge is found above the dark mask and passes above the tympanum to midbody on the sides. The dark gray dorsum often has a thin, black, middorsal stripe and/or a series of three light spots arranged in a triangular pattern. The venter is yellowish white. Some individuals have a dark bar on the top of the head between the eyes. The anterior surfaces of the limbs (from the elbow to the hand and from the knee to the ankle) have a distinct black stripe. The dorsal half of the iris is coppery, and the ventral half is black. No distinctly enlarged toe disks are present. The long hind limbs are used to escape predators with a series of long leaps. This species is most similar to Noble's Rainfrog *(E. noblei),* but the groin and posterior surface of the thigh of Noble's Rainfrog are red, as opposed to gray to white in the Northern Masked Rainfrog.

SEXUAL DIMORPHISM: The female is larger than the male. The tympanum is larger in the male (about the size of the eye) than in the female (smaller than the eye).

HABITAT: The Northern Masked Rainfrog is distributed throughout primary and secondary forests at La Selva, wherever leaf litter accumulates. The species is nocturnal and can be found by day under leaf litter and by night perched on low understory vegetation.

DIET: This frog is known to consume orthopterans, hemipterans, isopterans, lepidopteran larvae, isopods, and centipedes at La Selva (Lieberman 1986).

REPRODUCTION: We have not observed the male of this species to produce a call. Presumably, this species builds a ground nest. Juvenile abundance increases during the dry season, suggesting a late wet season bout of reproduction (Watling and Donnelly 2002).

REMARKS: The other common name Northern Mimicking Rainfrog is from Leenders (2001). However, it is not clear to us what this species is mimicking unless it is the Southern Masked Rainfrog *(E. gollmeri)*, a very similar species that is found in extreme southeastern Costa Rica. Those using our book in sites near the Panama border should compare frogs that key to the Northern Masked Rainfrog with the species description for the Southern Masked Rainfrog (see under "Additional Species").

NOBLE'S RAINFROG   *Eleutherodactylus noblei*
Pls. 55, 56

OTHER COMMON NAMES: Noble's Robber Frog

SIZE: Largest male 21 mm SVL (n = 1); largest female 47 mm SVL (n = 1). Leenders (2001) lists the adult male as being up to 53 mm and the adult female as being up to 66 mm.

DISTRIBUTION: Caribbean slopes from Honduras to Panama; Pacific slopes of southwestern Costa Rica and western Panama.

IDENTIFYING FEATURES: This species is reddish brown (juvenile) to gray (adult) in dorsal coloration. The groin and the back of the thigh are orange to red. In the juvenile, the coloring of the dorsum may be dissected by a network of gray lines; in the adult, the dorsum is uniform. Younger individuals have a relatively distinct, dark brown eye mask; this becomes indistinct in the adult. The iris of this species is golden. Noble's Rainfrog looks most similar to the Northern Masked Rainfrog *(E. mimus),* but the latter species lacks the red in the groin and on the back of the thigh and has greatly enlarged disks on the outer two fingers.

SEXUAL DIMORPHISM: The male is presumed to be smaller than the female. Also, the tympanum is larger in the male (as large as the eye) than in the female (smaller than the eye).

HABITAT: The species is found on leaf litter by day and perched low on understory vegetation at night. It is rare, relative to the other rainfrogs at La Selva that have a similar life style (Fitzinger's Rainfrog *[E. fitzingeri],* the Slim-fingered Rainfrog *[E. crassidigitus],* the Talamancan Rainfrog *[E. talamancae],* and the Northern Masked Rainfrog).

DIET: Presumably this frog consumes arthropods.

REPRODUCTION: We have not observed the male of this species to produce a call. Presumably this animal nests on the ground.

LOWLAND RAINFROG   *Eleutherodactylus ranoides*

Pl. 57

SIZE: Largest male 36 mm SVL (n = 8); largest female 67 mm SVL (n = 19).

DISTRIBUTION: Caribbean slopes from eastern Nicaragua to southeastern Costa Rica; Pacific slopes from Costa Rica to western Panama.

IDENTIFYING FEATURES: The dorsum of this frog is dark olive green and is warty relative to that of the other members of the genus at La Selva. The labial regions of the upper and lower jaws are dark gray to black with irregular, white to light yellow mottling. A thin, black stripe may extend from behind the eye, curving ventrally around the tympanum to the axilla (armpit). The venter is bright yellow. This species differs from all other members of the genus *Eleutherodactylus* at La Selva in possessing some webbing between the toes. However, this tissue covers only a small portion of the region between the toes and can be seen only by spreading the toes at their base. The species is also unusual among La Selva's rainfrogs in lacking expanded disks on the fingers.

SEXUAL DIMORPHISM: The male is smaller than the female. Additionally, the tympanum is larger in the male (about the size of the eye) than in the female (smaller than the eye).

HABITAT: This largest of La Selva's rainfrogs is a resident of stream- and riverbanks. It readily leaps into water when disturbed. The species can be found along the larger tributaries of the Rio Puerto Viejo.

DIET: Presumably this frog eats arthropods.

REPRODUCTION: We have not observed the male of this species produce a call. Presumably this frog digs a ground nest, perhaps under rocks along streams.

REMARKS: Older literature will refer to this species as *E. rugulosus.*

PYGMY RAINFROG
Pl. 58

Eleutherodactylus ridens

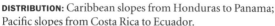

OTHER COMMON NAMES: Rio San Juan
Robber Frog
SIZE: Largest female 25 mm SVL (n = 9).
Leenders (2001) lists the male as being up to
19 mm SVL.
DISTRIBUTION: Caribbean slopes from Honduras to Panama;
Pacific slopes from Costa Rica to Ecuador.
IDENTIFYING FEATURES: The dorsal ground color is brown with a
faint, interorbital, dark bar and a faint, dark, W-shaped mark behind the head. Supraocular tubercles are present. A black stripe
extends from the nares to the eye and continues to the shoulder
in some individuals. The venter is white, and the chin is gray. The
groin, the anterior and posterior thigh, and the calf are red. The
iris is coppery brown on the dorsal half and dark brown on the
ventral half. Field characteristics that distinguish the Pygmy
Rainfrog from other species are the presence of expanded disks
on the outer two fingers, supraocular tubercles on the eyelids,
and a red wash in the groin and on the posterior thigh. This
species may be confused with the Clay-colored Rainfrog *(E.
cerasinus)* and the Golden-groined Rainfrog *(E. cruentus),* but
the Clay-colored Rainfrog lacks distinct supraocular tubercles
and the Golden-groined Rainfrog has a yellow gold wash in the
groin.
SEXUAL DIMORPHISM: We know of no external features that distinguish the two sexes.
HABITAT: This small, arboreal frog is often encountered near wet
areas such as the Research and Cantarana swamps. Laboratory
measurements suggest that activity peaks at dawn and dusk
(Winter 1987).
DIET: A variety of arthropods compose the diet, including formicids and other hymenopterans, adult and larval coleopterans,
homopterans, acarines, isopods, and nonspider arachnids, but
araneids are the most frequent diet item (Lieberman 1986).
REPRODUCTION: We have not observed the male of this species to
produce a call. Presumably this frog nests on leaves or tree trunks.

TALAMANCAN RAINFROG

Pls. 59, 60

Eleutherodactylus talamancae

OTHER COMMON NAMES: Almirante Robber Frog

SIZE: Largest male 33 mm SVL (n = 44); largest female 48 mm SVL (n = 35).

DISTRIBUTION: Caribbean slopes from Nicaragua to eastern Panama.

IDENTIFYING FEATURES: The dorsum of this frog either is mottled with subtle shades of light and dark brown or is nearly uniform. The most distinctive feature is a light yellow upper lip, which contrasts with a dark eye mask. In some individuals the light lip marking extends as a stripe from the tip of the snout along the sides of the body to the groin. In other individuals, the light upper-lip stripe is dissected by a few vague, gray bars, but even in these frogs the light appearance of the lip is a dominant feature. On the venter, this frog is white anteriorly shading to yellow posteriorly. The groin, posterior thigh, and ventral surfaces of the calf and heel are uniform purple to tomato red; the iris is copper with black flecks. This species can be distinguished from its close relatives at La Selva, the Slim-fingered Rainfrog *(E. crassidigitus)* and Fitzinger's Rainfrog *(E. fitzingeri),* by the light upper lip.

SEXUAL DIMORPHISM: The female is presumed to be much larger than the male. Additionally, the tympanum is larger in the male (about the size of the eye) than in the female (smaller than the eye).

HABITAT: This frog has elongate hind limbs for leaping and is found at night perched low on vegetation in primary forest. Activity peaks at dawn and dusk (Winter 1987).

DIET: This frog consumes arthropods (formicids, dipterans, coleopteran larvae, isopods, and nonspider arachnids), as well as occasional vertebrates; however, araneids and orthopterans are the major diet items at La Selva (Lieberman 1986).

REPRODUCTION: The male produces a high-pitched, single-note call that sounds like a cat "mew" and is given at widely spaced intervals. Presumably this species builds nests on the ground. Juvenile abundance increases during the dry season, suggesting a late wet season bout of reproduction (Watling and Donnelly 2002).

FRINGE-TOED FOAMFROG   *Leptodactylus melanonotus*
Pls. 61, 62

OTHER COMMON NAMES: Black-backed Frog,
Sabinal Frog

SIZE: Largest adult 44 mm SVL (n = 20).

DISTRIBUTION: Caribbean slopes from Tamaulipas,
Mexico, to northern Colombia; Pacific slopes from
Sonora, Mexico, to Ecuador.

IDENTIFYING FEATURES: This drab, medium-sized frog is found exclusively in or near standing water. The adult is dark olive, with few distinguishing features. A series of warts on the dorsum forms a weak, dorsolateral ridge on the posterior half of the body. The juvenile is slightly more colorful than the adult, being mottled with dark reddish brown and black. The juvenile also has a faint, black triangle on top of the head between the eyes, with the apex of the triangle extending posteriorly. All individuals have dark lips with two to three white spots on each side, reddish arms, black posterior thigh surfaces with whitish gray spots, and a white venter with dark gray mottling. The species can be told from ranids of similar size by the restricted extent of toe webbing (ranids have webs extending at least halfway up each toe). The only species of frog at La Selva with which the Fringe-toed Foamfrog might be confused is the Lowland Rainfrog *(Eleutherodactylus ranoides);* the latter has a yellow, rather than white, venter.

SEXUAL DIMORPHISM: The male develops two pointed, black, keratinized spines at the base of each thumb during the breeding season. These are thought to improve the grip of the male during amplexus.

HABITAT: This species occurs at both the Cantarana and Research swamps at La Selva. Additionally, the male can be heard calling from flooded fields, roadside ditches, and swampy areas along the Rio Puerto Viejo and the Suampo trail.

DIET: This frog is presumed to feed on arthropods.

REPRODUCTION: The adult male produces a soft "guh" or metallic "doink" that, when the frog is in large choruses, is given as a series of repeated notes at roughly one-second intervals. The male calls from leaves floating on the water surface. The calling sites are usually hidden from view, so this frog is often difficult to locate. An amplectant pair creates a floating foam nest, within which eggs are deposited and tadpoles begin to develop. The foam is produced by anal glands and is consumed by developing tadpoles early in their lives. Breeding takes place in the wet season.

REMARKS: Individuals that key to the Fringe-toed Foamfrog should be compared with the description of the Turbo White-lipped Foamfrog *(Leptodactylus poecilochilus)* (see under "Additional Species").

SMOKY JUNGLE FROG   *Leptodactylus pentadactylus*

Pls. 63–65

OTHER COMMON NAMES: Rana Comepollos,
Rana Ternero, Central American Bullfrog,
South American Bullfrog

SIZE: Largest male 141 mm SVL (n = 19);
largest female 136 mm SVL (n = 15). Leenders
(2001) lists the male as being up to 169 mm and the
female as being up to 181 mm.

DISTRIBUTION: Caribbean slopes from Honduras to the Amazon Basin of northern Brazil; Pacific slopes from Nicaragua to Ecuador.

IDENTIFYING FEATURES: The adult of this large, distinctive species has a dorsal reticulum of dark purplish and light brown. A dark, interorbital bar is always present. The lips are boldly marked with light and dark bars. The venter is gray with white punctations, and the posterior thigh is black with small white spots. The iris is cinnamon brown. The juvenile is more boldly marked, with a wide, reddish brown area between weak dorsolateral folds and a dark stripe from the nostril to the eye. The only species likely to be confused with this one is Vaillant's Frog *(Rana vaillanti)*, which is smaller in size, is less colorful, and has extensive webbing between the hind toes.

SEXUAL DIMORPHISM: The female is slightly larger than the male. The adult male can be distinguished from the female by the presence of a black, cornified spine projecting from the base of each thumb and from the right and left sides of the chest. These are thought to improve the grip of the male during amplexus.

HABITAT: This species is a denizen of forests near slow-moving streams and open swamps. The adult uses a burrow located in loose soil and, at night, is often found positioned at the burrow entrance, where it can be located by "eye shine" (red light reflected off the tapetum lucidum). The adult male calls from burrow entrances or from swamps.

DIET: At La Selva this frog is known to consume formicids and coleopterans (Lieberman 1986). Elsewhere, it eats a variety of prey including large invertebrates and small vertebrates, especially frogs using temporary pools (Scott 1983c).

REPRODUCTION: Reproduction occurs during the first major rains of the rainy season (Blankenship 1993; Galatti 1992; Muedeking and Heyer 1976; Rivero and Esteves 1969). At this time the male gives a low, resounding "whoop" call; we have heard it principally near the Research and Cantarana swamps. Males may actively defend positions in the choruses (Rivero and Esteves 1969). An amplectant pair creates a foam nest from cloacal secretions, water, and semen, and the female deposits about 1,000 fertilized eggs in it. The nest is sometimes placed in water or on land near areas that will flood (Blankenship 1993; Muedeking and Heyer 1976; Rivero and Esteves 1969; Vinton 1951). Tadpoles hatch and spend some time in the nest consuming foam before entering the water column or transforming (Vinton 1951). Although the tadpoles of this species are known to consume other tadpoles, specimens in the laboratory grow fastest on a diet of vegetation (Blankenship 1993). Offspring are produced over relatively short time periods early in the wet season, creating distinct yearly cohorts that reach maturity the following year (Galatti 1992). However, these cohorts can include individuals from more than one egg-laying event, each associated with intense rain showers spaced as much as one month apart (Blankenship 1993).

REMARKS: When captured, the adult often gives a prolonged, high-pitched scream, reminiscent of a crying baby. This call has been presumed to be a warning to conspecifics, but no behavioral data on potential recipients has been collected, and caiman are thought to orient toward this call (Scott 1983c). The adult possesses potent skin toxins that can kill other frogs placed in close proximity (in the same collecting bag, for example) and that can cause a rapid allergic response in sensitized humans (causing a runny nose, puffy eyes, and sneezing).

Microhylidae

This family contains approximately 320 species that are distributed in the New World from the southeastern United States, through Mexico and Central America, to Amazonian South America. In the Old World, microhylids are found in equatorial and southern Africa, in eastern Asia to extreme northern Australia, and on New Guinea. These frogs have invaded a variety of terrestrial and arboreal habitats and can withstand a variety of environmental conditions, including those of desert regions. In the lowland wet forests of Costa Rica, these anurans are called narrow-mouthed toads because the heads of the Neotropical members of this family are small and triangular, leaving little room for a mouth. These animals are also referred to as sheep frogs because their call sounds like a bleating sheep. Only one representative of this family is present at La Selva, and it is the only one known to occur in the Caribbean lowland wet forests of Costa Rica.

RETICULATED SHEEPFROG *Gastrophryne pictiventris*
Pls. 66, 67

OTHER COMMON NAMES: Sheep Frog, Narrow-mouthed Toad, Nicaragua Narrowmouth Toad

SIZE: Largest male 35 mm SVL (n = 11); largest female 37.5 mm SVL (n = 70).

DISTRIBUTION: Caribbean slopes of Nicaragua and Costa Rica.

IDENTIFYING FEATURES: The Reticulated Sheepfrog is uniform dark gray above or gray with a pair of irregular, dark gray streaks down the back. A wide, black mask, bordered above by a thin white or yellow line, extends from the tip of the snout, through the eye, to the level of the groin. The iris is reddish gold with a black reticulum. The venter has a reticulum of white spots or blotches on a dark gray to black background. A narrow fold of skin across the back of the head is a good field characteristic for the genus, but you need a hand lens and/or sufficient light to see it. Additionally, the toadlike body (small triangular head, fat body, and short hind legs) and the absence of parotoid glands distinguish this species from all other anurans at La Selva.

SEXUAL DIMORPHISM: Presumably the female is larger than the male.

HABITAT: Until pitfall trapping was performed at La Selva (Lieberman 1986), this species was thought to be rare. It is, in fact, relatively abundant in the leaf litter, but its secretive, semifossorial habits make it difficult to find. The species appears to be fairly common in damp areas near streams that flood.

DIET: The adult has a relatively specialized diet of isopterans and large ponerine ants; it is also known to eat coleopterans, orthopterans, non-ant hymenopterans, dipterans, hemipterans, araneids, nonspider arachnids, centipedes, acarines, and oligochetes (Lieberman 1986).

REPRODUCTION: The male attracts females with a nasal, sheeplike call, a prolonged "whaaaa," given from small, temporary pools in the leaf litter environment of recent alluvium. Calling can take place day or night, and multiple males often call in unison for five to 10 seconds followed by five to 10 seconds of silence. Eggs are laid as a surface film. The tadpoles are readily distinguished from those of other frog species at La Selva because they lack keratinized mouthparts (Donnelly, de Sá, et al. 1990). Instead, they possess mouthparts designed to feed on microscopic crustaceans captured by protrusion of the lower jaw. The pools can dry quickly; the one from which the tadpoles described by Donnelly, de Sá, et al. (1990) were taken retained water for only 18 days. Juvenile abundance increases during the dry season, suggesting a late wet season bout of reproduction (Watling and Donnelly 2002).

Dendrobatidae

These are the dart-poison frogs, so named because some native cultures in South America use three members of the family as a source of poison for blowgun darts. The family contains approximately 200 species that are distributed from the wet forests of Nicaragua to southeastern Brazil. These frogs are easily distinguished by their bright, aposematic coloration and by their well-developed system of parental care. The bright colors are associated with skin toxins composed of a variety of alkaloids, an unusual class of biological chemicals. Dart-poison frogs do not synthesize these chemicals but instead sequester them from arthropods (Daly et al. 1994). Females of all species within the family Dendrobatidae lay eggs in terrestrial nests. Typically, both sexes remain with the nest until the eggs hatch. Then one or both parents carry the tadpoles on their backs to an aquatic site, where tadpole growth takes place. Females of some species revisit the developing young and feed them unfertilized eggs. The three species at La Selva are small and diurnal, and they represent the fauna known from the rest of the Caribbean lowland wet forests of Costa Rica as well.

Key to the Dendrobatidae at La Selva

1a   Body bright red to orange with dark blue to black legs (rarely, body uniform blue with black spots)..............
.... *Dendrobates pumilio*
1b   Body does not have red as the principal color............2
  2a   Body black with green blotches or winding stripes ...
.... *Dendrobates auratus*
  2b   Body black with a pair of yellow, yellow orange, or turquoise, dorsolateral stripes.... *Phyllobates lugubris*

GREEN AND BLACK POISON FROG   *Dendrobates auratus*
Pl. 68

OTHER COMMON NAMES: Green Poison Frog, Green and Black Dart-poison Frog, Gold Arrow-poison Frog

SIZE: We have no size data from La Selva, but elsewhere this species ranges up to 42 mm SVL.

DISTRIBUTION: Caribbean slopes from southern Nicaragua to extreme eastern Panama; Pacific slopes from the Osa Peninsula, Costa Rica, to Colombia.

IDENTIFYING FEATURES: The adult of this species is glossy dark brown to black with broad blue green anastomosing stripes. The venter is marbled or spotted with yellow, blue, or green on a dark background. The iris is black. No other frog at La Selva can be confused with this one.

SEXUAL DIMORPHISM: The male is smaller and thinner than the female.

HABITAT: This frog is found in primary and secondary forest, typically on leaf litter and often associated with streams. It is diurnal in activity.

DIET: The diet of this species is unstudied at La Selva; elsewhere this frog consumes formicids, coleopterans, lepidopterans, chilopods, acarines, and collembolans (Silverstone 1975; Toft 1981).

REPRODUCTION: The male produces a repeated buzzing call of three to five notes ("cheez-cheez-cheez;" Savage 1968) that is used to attract females. A male that is selected by a female then leads the female to a nest site, under leaf litter, where four to six eggs are deposited and fertilized (Kitasako 1967). The male then transports the tadpoles one at a time to temporary pools of water (Eaton 1941), where they take 68 to 102 days to develop (Silverstone 1975).

REMARKS: This species was introduced to the Selva Verde Lodge in Chilamate in 1986 by a local worker. It has flourished there ever since. The species was first observed at La Selva in 1996 by Danilo Brenes, who found one at the Huertos plots (around STR 1200). The species either crossed the Rio Sarapiquí or was assisted by a local worker.

STRAWBERRY POISON FROG   *Dendrobates pumilio*
Pl. 69

OTHER COMMON NAMES: Ranita Roja, Blue Jeans Frog, Flaming Poison Frog, Red-and-blue Poison Frog, Strawberry Dart-poison Frog

SIZE: Largest male 22 mm SVL (n = 150); largest female 22 mm SVL (n = 243).

DISTRIBUTION: Caribbean slopes from Nicaragua to western Panama.

IDENTIFYING FEATURES: This boldly marked animal is unusual among La Selva's frogs in its diurnal activity pattern and ubiquitous distribution. Because of its bright colors, the Strawberry Poison Frog cannot be confused with any other frog species. The dorsum is orangish red, often with numerous small, black spots. The red color changes abruptly to blue black on the limbs. In rare instances, this frog is uniform dark blue with black spots. The iris is jet black. When recently transformed, the frog is uniform maroon.

SEXUAL DIMORPHISM: The sexes are similar in size, but the mature male has a brown throat patch whereas the throat of the female is red (Donnelly 1989c).

HABITAT: This species is widely distributed at La Selva, occurring in any habitat where trees form a more or less continuous canopy. It is particularly abundant (or easy to see) in the Arboretum and cacao groves *(Theobroma cacao)*. Activity is largely diurnal.

DIET: This animal is a gape-limited predator specializing on formicids and acarines; it is also known to consume coleopterans, orthopterans, dipterans, homopterans, collemoblans, isopods, diplopods, and araneids (Lieberman 1986; Limerick 1976). An ontogenetic shift in diet occurs, with the juvenile consuming mostly acarines and the adult, especially the female, switching to formicids (Donnelly 1991).

REPRODUCTION: The call of the male is a repetitive "eeh-eeh-eeh… " that is reminiscent of the sounds created by katydids. The male is territorial. The male emits advertisement calls during the day to attract mates and to inform neighboring males of its location. The male defends its territory by wrestling with intruders (Robakiewicz 1992). Such contests are typically won by the resident male (Baugh and Forester 1994). Courtship is prolonged (Limerick 1976), and a pair deposits eggs and sperm in a terrestrial nest located under leaf litter. This species is unusual in having biparental care. The male tends the eggs, and the female tends the tadpoles. The male is known to moisten the clutch with water released from his bladder (Weygoldt 1980) and may defend the clutch from other males. After hatching, tadpoles wiggle onto the female parent's back and are deposited in the water that accumulates in the axils of the leaves of bromeliads and aroids. The female parent repeatedly visits the offspring, depositing unfertilized eggs that serve as food for the developing larvae (Weygoldt 1980). Population density peaks during the early wet season, suggesting increased recruitment associated with moisture accumulation during that time (Donnelly 1989a, 1989c).

REMARKS: Individuals have been observed to orient and return to their original territories after being displaced 20 m from home (McVey et al. 1981).

STRIPED DART-POISON FROG   *Phyllobates lugubris*
Pls. 70, 71
OTHER COMMON NAMES: Lovely Poison Frog
SIZE: Largest adult 23 mm SVL (n = 10).
DISTRIBUTION: Caribbean slopes from
Nicaragua to western Panama.
IDENTIFYING FEATURES: This frog is boldly marked with a jet black dorsum and a pair of wide, yellow or turquoise dorsolateral stripes. The hind legs are black with a turquoise reticulum. A thin, light turquoise stripe extends from the tip of the

snout, below the eye, along the upper lip, to the axilla and extends down the dorsal surface of the arm. The venter is black with a turquoise reticulum; this coloring extends along the sides as well. The front and back of the thighs are black with turquoise spots. The iris is dark brown to jet black, making the eye difficult to distinguish against the black lateral portion of the head and body. No other frog at La Selva can be confused with this one.

SEXUAL DIMORPHISM: We know of no external features that distinguish the two sexes. Mean sizes given by Silverstone (1976) indicate that the female is slightly larger than the male.

HABITAT: The Striped Dart-poison Frog is uncommon at La Selva, but it has been found in litter samples taken in both primary and secondary forest. This species is heard most consistently in a low area along the SOR trail near the entrance to Rafael's house.

DIET: This frog consumes formicids (Lieberman 1986).

REPRODUCTION: The male produces a trilled whistle, somewhat reminiscent of a buzzing insect. This call is used to establish a territory and to attract mates. The male and female create a ground nest within which the female deposits eggs and the male fertilizes them. The male then transports the tadpoles to unknown aquatic rearing sites. This is done by carrying five to 10 tadpoles at a time (Donnelly, Guyer, et al. 1990). Caldwell (1994) observed adults of this species transporting tadpoles to a small pool near one used by the Grey-eyed Leaf Frog *(Agalychnis calcarifer)*. Little else is known regarding parental care.

REMARKS: Three members of this genus are used by South American human cultures to create poison-tipped darts. This is done by rubbing the darts across the back of a live frog, thereby transferring the alkaloid toxins produced by the frog to the darts, which are then used with blowguns. Hence, this genus includes the true dart-poison frogs. However, the toxins of the Striped Dart-poison Frog are not particularly potent, and no human cultures within the range of this species have used this frog to create poisonous darts. The other common name for dart-poison frogs is poison-arrow frogs, but no human culture has used the members of the genus *Phyllobates* in this manner (Myers et al. 1978). One species from the family Leptodactylidae, Gaige's Rainfrog *(Eleutherodactylus gaigeae),* mimics the Striped Dart-poison Frog in color pattern. This mimic is found in extreme southeastern Costa Rica. Therefore, if you are using our book for animals captured near the Panama border, you should compare individuals that key to the Striped Dart-poison Frog with the description of Gaige's Rainfrog (see under "Additional Species").

Ranidae

The ranid frogs comprise approximately 700 species that are widely distributed, being found on all major continents except Antarctica. Unfortunately, there is no diagnostic feature for the family, which probably means that the members of this family are not all one another's closest relatives. Most of these frogs are long-legged and live near water. However, a few genera have toad-like bodies and burrow, and others have brightly colored skin and alkaloid skin toxins, like members of the family Dendrobatidae. Still others have invaded arboreal habitats. Many ranids breed in aquatic habitats and have aquatic tadpoles. However, direct development occurs in some members of the family. In the lowland wet forests of Costa Rica, ranids are characterized by smooth skin, elongate hind limbs that are used for leaping, fingers and toes that are narrow and lack disks, and extensive webbing on the hind feet. The La Selva herpetofauna includes two species of ranid frogs. One additional species is known from elsewhere in the Caribbean lowland wet forests of Costa Rica.

Key to the Ranidae at La Selva

1a One to three bold, round, yellow spots on posterior surface of thigh. . . . . *Rana warszewitschii*

1b No bold yellow spots on posterior surface of thigh (yellow mottling may be present). . . . . *Rana vaillanti*

VAILLANT'S FROG   *Rana vaillanti*

Pls. 72, 73

SIZE: Largest male 81 mm SVL (n = 3); largest female 102 mm SVL (n = 6).

DISTRIBUTION: Caribbean slopes from Vera-cruz, Mexico, to northern Colombia; Pacific slopes from Oaxaca, Mexico, to Ecuador.

IDENTIFYING FEATURES: The adult has a tannish brown dorsum with a greenish cast, often with small, black punctations. A pair of dorsolateral ridges is present. The top of the head and sides of face are green. The upper lip is uniform grayish green, and the lower lip is white or gray, extending as a stripe to the anterior portion of the upper arm. The venter is white to light yellow, and the posterior thighs are dark gray with a yellow reticulum. The dorsal quarter of the iris is yellow, and the ventral three-quarters are mottled with light and dark gray. The juvenile is dark green with a wide, tan stripe between the dorsolateral ridges. This region frequently has small, black punctations. The sides of the body of the juvenile are brownish gray, and the lips are yellowish. The side of the face is green bordered dorsally by a distinct black line from the naris to the eye. This is the only frog at La Selva with these colors, extensive toe webbing, and thin toes lacking expanded disks.

SEXUAL DIMORPHISM: The female is larger than the male. Additionally, the reproductive male has a thickened and darkened pad along the inside (medial surface) of each thumb and thicker forearms than the female.

HABITAT: This drab frog is the most common ranid at La Selva. It occupies deeper pools of stagnant or slow-moving water. This frog is common at the Cantarana and Research swamps, where it can be encountered by day or night on the boardwalk, on logs or floating vegetation, or at the water's surface.

DIET: Like many ranids, this frog consumes a wide variety of prey including small vertebrates (even conspecifics; Noble 1918). However, most diet items are insects (principally coleopterans and odonates, the latter captured on their night roosts) and araneids (Ramirez et al. 1997).

REPRODUCTION: The call of the adult is an incredible complex of squalls and chortles, giving the impression that several species are present. The call is reminiscent of the sounds produced by rubbing a hand across an inflated balloon. Calls are given from water covered by thick vegetation. Nothing is known of the timing of reproduction or of associated behaviors. Eggs, about 1,000 per clutch, are deposited in large pools during the wet season.

REMARKS: A frightened Vaillant's Frog gives a high-pitched, single-note "yip" call while leaping toward safety, and a captured individual may give a prolonged, high-pitched call when handled. Most previous species lists refer to this frog as *R. palmipes*. We follow the systematic treatment of Hillis and de Sá (1988) in calling the Central American form *R. vaillanti*. The call given by this frog at La Selva is similar to that of Taylor's Leopard Frog (formerly *R. pipiens,* now *R. taylori*) and its relatives. The inclusion of Taylor's Leopard Frog on earlier species lists for La Selva (Scott et al. 1983) may have resulted from a misidentification of the call of Vaillant's Frog. However, we include a description of Taylor's Leopard Frog (see under "Additional Species") because this species is known from lowland sites near La Selva and, therefore, may be present on the property.

BRILLIANT FOREST FROG   *Rana warszewitschii*

Pls. 74–76

OTHER COMMON NAMES: Warszewitsch's Frog
SIZE: Largest male 43 mm SVL (n = 5); largest female 65 mm SVL (n = 10).
DISTRIBUTION: Caribbean slopes from Honduras to Panama; Pacific slopes from Costa Rica to eastern Panama.

IDENTIFYING FEATURES: Unlike most ranids, this is an attractive frog. Well-developed, dorsolateral ridges extend the length of the body, separating a brown middorsal region from a lateral region that is black or red. The middorsum may be suffused with small, green flecks or spots. A wide, black mask extends from the tip of the snout, through the eye, to the level of the groin. The iris is bronze to dark brown and is generally difficult to distinguish from the black mask. The lips are uniform cream to yellow, especially posterior to the eye, and the venter is yellow, sometimes mottled with red. The posterior surface of the thigh is black with one to three large, yellow spots; the black on the thigh shades to tomato red ventrally. This species could be confused with members of the genus *Eleutherodactylus,* but the Brilliant Forest Frog is the only forest-dwelling frog with fully webbed feet and no expanded toe discs.

SEXUAL DIMORPHISM: The female is larger than the male.

HABITAT: This species can be found by day or by night on leaf litter in primary forest, where it wanders far from permanent water. We have seen this frog frequently at night during the wet season along the Camino Circular.

DIET: Presumably this frog eats arthropods.

REPRODUCTION: The reproductive biology of this species is virtually unknown. Caldwell (1994) collected tadpoles in pools and small streams during the dry season. This species does vocalize, even though no vocal sacs or slits are present. The call has been described as a short trill (Greding 1972).

REMARKS: We follow Hillis and de Sá (1988) for the spelling of the specific epithet; previous authors have used the spelling *R. war-schewitschii.*

REPTILIA

REPTILES ARE LAND vertebrates that have epidermal scales composed of beta keratin, eyes designed for rapid focusing, color vision, and amniotic eggs (these are usually shelled eggs with associated embryonic membranes but are sometimes unshelled eggs that implant on the uterus). The scales that cover the body are usually thick enough to create an effective barrier to water loss and at the same time thin enough to allow the acquisition of heat from the environment. About 17,200 living species of animals are derived from an ancestral reptile that first appeared in the fossil record in the mid-Carboniferous Period (about 350 million years before present). The vast majority of these species are either birds (10,000 species) or squamates (lizards and snakes; 6,900 species). Thus, the group to which all modern reptiles belong is the most successful group of land vertebrates, if success is defined in terms of species richness.

Reptiles have internal fertilization, a process that requires that males exhibit stereotypical courtship behaviors toward potential mates. Accumulating evidence suggests that females of all reptile orders actively select among courting males, making choices based on body size and the vigor with which males display. Once inseminated, females either deposit shelled eggs in a nest or retain the eggs in the uterine tract and produce live young. Reptile species differ greatly in the level of parental care provided. Some provide no care beyond creation of a nest; in others, both sexes attend the nest, assist the young in exiting the nest, and interact with newborn offspring during early development.

Living reptiles include species with widely divergent foraging methods. For this reason reptiles perform a wide variety of ecological functions. Some forms are herbivorous and play an important role in cropping understory vegetation. Others consume fruits and, therefore, disperse the seeds of flowering plants. Many species are predators, consuming prey as small and abundant as ants and as large and relatively rare as medium-sized mammals. Some species sit and wait for prey to approach, whereas others actively search for prey. Additionally, many species of reptiles are abundant and, therefore, are the prized prey of other predators.

Key to the Orders of Reptiles at La Selva

1a   Body encased in a shell .... Testudines
1b   Body not encased in a shell 2
     2a   Tail thick, muscular, and laterally compressed for swimming .... Crocodylia
     2b   Tail variable, but if thick and muscular then not laterally compressed for swimming .... Squamata

TESTUDINES

This order contains the 260 species of living turtles. Turtles are distinctive reptiles because of the protective shell that covers their bodies. This unique structure is associated with a number of other unusual features. Because the upper part of the shell is composed, in part, of fused ribs and vertebrae, turtles are unable to breathe via expansion of the rib cage, a process used by all other reptiles. Instead, turtles expand the neck region using specialized muscles, thereby ventilating the lungs. Turtles also have S-shaped limb bones, which allow them to crawl with the shell positioned off the ground, and a unique joint where the neck joins the torso that allows the head to be retracted into or alongside the shell. Finally, turtles are unusual in that they lack teeth.

All turtles lay shelled eggs and, therefore, have internal fertilization. Copulation occurs via the insertion of the male's erectile penis into the cloacal opening of the female. Males must actively court females in order to be selected for mating. Courting behaviors frequently involve males scratching or nibbling the head, neck, and anterior portion of the shell of females. Once inseminated, the female turtle selects the nest site. Most nests are dug in loose soil with the hind feet. Eggs are deposited in the nest, covered over, and left to develop. However, in some forest-dwelling turtles, eggs are simply laid on or under leaf litter. Development inside the egg typically takes 50 to 60 days; however, eggs laid late in the year may develop more slowly and hatch early in the following year. In some turtles, the sex of the individual is determined by sex chromosomes. In others, sex is determined by the temperature of the nest within which the individual develops.

Turtles eat a variety of foods. Some are herbivorous, consuming the leaves, flowers, and fruits of low-lying vegetation. Others are predators, eating a variety of aquatic invertebrates and fish. Few predators can consume adult turtles. However, eggs and juveniles are prized prey items for carnivorous mammals, birds, and snakes. The protection from predation provided by the shell of the adult may explain why, among vertebrates, turtles have unusually long life spans. Most species of turtle are likely to live as long as or longer than humans.

Most turtles of the lowland wet forests of Central America are found in slow-moving or stagnant waters. Members of the family Emydidae frequently bask on logs in rivers or on the ground in clearings along riverbanks and, therefore, can be seen if you float quietly down a river. Members of the other two families, Chelydridae and Kinosternidae, rarely appear on land, except when searching for nest sites or new aquatic environments. Chelydrids are large animals that spend most of their time at the bottom of moving waters that are sufficiently shallow to allow them to extend their necks and reach the water's surface to breathe. They can be found by searching river bottoms with a mask and snorkel. Kinosternids inhabit the margins of stagnant pools of water and slow-moving streams. They can occasionally be observed by surveying appropriate habitat. Funnel traps baited with fish or chicken are often successful in capturing turtles. Whenever possible, turtles should be released immediately and should be released back where they were captured. If individuals are retained, they can be kept in moistened cloth bags for short periods of time, but no more than a few days. Although turtles lack teeth, their jaws have sharp, slicing edges, and all species are capable of delivering painful bites. Therefore, anyone choosing to handle turtles should do so carefully.

Key to the Families of Testudines at La Selva

1a	Plastron (ventral portion of shell) with less than 12 scutes (fig. 10a, b)..2	
1b	Plastron with 12 scutes (fig. 10c)...............Emydidae	
	2a	Plastron extremely reduced, cruciform (fig. 10a); tail long, more than half of body length.....Chelydridae
	2b	Plastron not cruciform (fig. 10b); tail short, less than half of body length...................Kinosternidae

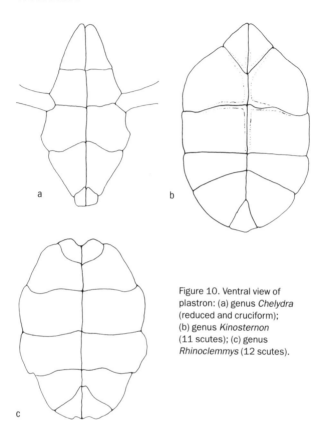

Figure 10. Ventral view of plastron: (a) genus *Chelydra* (reduced and cruciform); (b) genus *Kinosternon* (11 scutes); (c) genus *Rhinoclemmys* (12 scutes).

Chelydridae

Members of this family are referred to as snapping turtles, because they are large with enormous heads and, therefore, can deliver a snapping bite. The family contains three species distributed in the eastern half of the United States and from extreme southern Mexico to northern South America. These turtles live in slow-moving freshwater rivers and eat a variety of animal and plant foods. Only one species of this family of turtles is found at La Selva, and it is the only species known to occur in the Caribbean lowland wet forests of Costa Rica. The characteristics given in the species account distinguish this family from the other two families of La Selva's turtle fauna.

COMMON SNAPPING TURTLE *Chelydra serpentina*
Pl. 77

OTHER COMMON NAMES: Tortuga Lagarto, Talmana

SIZE: No specimen from La Selva has been measured; elsewhere this species reaches ca. 500 mm CL (Acuna 1993).

DISTRIBUTION: Eastern North America to the Rio Grande drainage of Texas; Caribbean drainages from Veracruz, Mexico, to Panama; Pacific drainages from Costa Rica to Ecuador.

IDENTIFYING FEATURES: This is the largest of La Selva's turtles. The body of this species is distinctive and cannot be confused with any other turtle. The carapace is dark (almost black in small individuals) and extremely rugose, with three longitudinal ridges along the carapacial scutes, but these become smooth and less noticeable in large adults. The plastron is reduced to a cross-shaped structure that exposes much flabby skin in the axilla and groin. The head is quite large and dark, with a strongly hooked upper jaw. Relative to other turtles, the tail is long, longer than half the length of the carapace. No other turtle at La Selva looks like this one.

SEXUAL DIMORPHISM: The male is much larger than the female and has a longer tail.

HABITAT: This species resides in deep waters and, therefore, is seen infrequently. At La Selva, the adult probably occurs in both major rivers, as well as in large creeks. This species has been observed in the Rio Puerto Viejo near the steps to the River Station, in the Quebrada Sura (at the Arboretum), and in the Quebrada Leone.

DIET: This turtle is a sit-and-wait predator, using its oddly shaped carapace to look like a fallen log. However, dietary analysis indicates omnivory, with the diet including both animal and plant material.

REPRODUCTION: A mating pair has been observed at Quebrada Leone (O. Vargas 1998, personal communication). Other than this observation, nothing is known about the reproduction of this species at La Selva; elsewhere in Costa Rica this turtle breeds from April to November (Acuna 1993). Early in the wet season, the female excavates a nest in the margin of a river, where 11 to 83 eggs are laid (Acuna 1993).

REMARKS: Care should be taken when handling this animal; pick it up by the base of the tail or the back of the carapace and hold it away from your body. The Common Snapping Turtle is not known to bite underwater, but when brought on land it uses its deceptively long neck and powerful jaws to lunge at and bite captors. This species is exploited for its meat (used in soups) and its eggs (used as counterfeit sea turtle eggs) (Acuna 1993).

Kinosternidae

This is the family of the mud turtles and musk turtles, so named for their preference for the muddy edges of marshes, swamps, slow rivers, and streams, and for the presence of a musk gland that exudes an odiferous compound onto the part of the shell between the legs where the carapace joins the plastron. The family is composed of approximately 22 species that are found from the eastern half of the United States, through Mexico and Central America, to Amazonian South America. A disjunct population of one species is found in northern Argentina. In the lowland wet forests of Costa Rica, these turtles are found in slow-moving rivers and streams, marshes, and open swamps. Two species of kinosternid turtles are present at La Selva, and they represent the fauna known from the rest of the Caribbean lowland wet forests of Costa Rica as well.

Key to the Kinosternidae at La Selva

1a   Plastron large, more or less completely covering venter when closed; venter not noticeably fleshy. . . . . *Kinosternon leucostomum*

1b   Plastron small, not covering venter when closed; venter extremely fleshy . . . . *Kinosternon angustipons*

NARROW-BRIDGED MUD TURTLE
Pl. 78

Kinosternon
angustipons

OTHER COMMON NAMES: Tortuga Pequeña,
Pecho Quebrado, Pecho en Cruz
SIZE: Largest male 107 mm CL (n = 1).
DISTRIBUTION: Coastal Caribbean drainages
from Nicaragua to northern Panama.
IDENTIFYING FEATURES: This species is uniform tan to
brown in dorsal coloration. It can be confused only with the
White-lipped Mud Turtle *(Kinosternon leucostomum)*, from
which it can be distinguished by a small plastron that does not
cover the entire venter. The Narrow-bridged Mud Turtle cannot
retract its limbs and close the plastron tightly against the carapace. The reduced plastron exposes large portions of flabby skin
when viewed from below.

SEXUAL DIMORPHISM: The male can be distinguished from the female based on the length of the tail. When stretched, the cloacal
opening is beyond the edge of the carapace in the male and is at
most to the edge of the carapace in the female. Additionally, the
male possesses a patch of spiny skin on the posterior surface of
the thigh and shank.

HABITAT: This is a rare turtle at La Selva and throughout its range.
The species is known only from a handful of sites along the Atlantic versant in lower Central America (Iverson 1992). It is the
smallest of La Selva's turtles and is known only from the Research
swamp but may also occur in the Cantarana swamp and perhaps
along the El Suampo trail.

DIET: This species is omnivorous; it consumes plants and animals
(Acuna 1993).

REPRODUCTION: Although no nest has been observed at La Selva,
nesting is thought to occur from May through August, when one
to four (usually four) eggs are placed in a small hole dug by the female (Acuna 1993).

WHITE-LIPPED MUD TURTLE   *Kinosternon leucostomum*

Pls. 79–81

OTHER COMMON NAMES: Polchitoque,
Tapaculo, Candaso Pequeño, Tortuga
Amarilla

SIZE: Largest male 157 mm CL (n = 11);
largest female 127 mm CL (n = 6).

DISTRIBUTION: Caribbean drainages from Veracruz,
Mexico, through Panama, to northern Colombia; Pacific
drainages from Costa Rica to Ecuador.

IDENTIFYING FEATURES: This species gets its scientific name from
the wide, light (usually yellowish) stripes on the upper and lower
jaws and cheeks. Sometimes these have a reddish cast in the region of the head behind the eyes. The iris is yellow with cross-
shaped, brown areas. A hinge is present on the plastron, and the
animal can completely withdraw the body into its shell and seal
the opening by closing the plastron against the carapace. This
feature distinguishes the White-lipped Mud Turtle from the
Narrow-bridged Mud Turtle *(K. angustipons).*

SEXUAL DIMORPHISM: The male is larger than the female in this
species. The tail of the male is longer and has a thicker, more
swollen base than that of the female. A handy field characteristic
can be seen by examining the position of the tail in ventral view:
if the tail reaches the level of the knee when wrapped under the
carapace, it is a male; if not, it is a female. Additionally, the male
possesses a patch of spiny skin on the posterior surface of the
thigh and shank. Morales-Verdeja and Vogt (1997) reported similar sexually dimorphic characteristics at Los Tuxtlas.

HABITAT: The White-lipped Mud Turtle inhabits muddy swamps
and slow-moving creeks. During the dry season, the turtle buries
itself and becomes quiescent (Acuna 1993). It leaves its burrow
during the first rains of the rainy season and migrates over land
to aquatic areas like the Research and Cantarana swamps. This
turtle is abundant at these sites and can be trapped relatively easily during the wet season in funnel traps baited with meat scraps.

DIET: This species is omnivorous; fecal matter from individuals
that we trapped contained a variety of aquatic insects. This turtle
also scavenges for dead animal and plant material and may disperse the seeds of some wetland plants (Acuna 1993; Villa 1973;
Vogt and Guzman 1988).

REPRODUCTION: In Costa Rica, the eggs of this species are laid between July and October and number one to four (usually one; Acuna 1993). However, the nesting season of this species differs elsewhere in its range (mid-August through the end of May at Los Tuxtlas; September through December in Tabasco, Mexico; Morales-Verdeja and Vogt 1997). The eggs are deposited on the ground under leaf litter (Acuna 1993), and this activity is usually nocturnal or crepuscular (Morales-Verdeja and Vogt 1997). Incubation temperature determines the gender of hatchlings in this species; nest temperatures between 25 and 27 degrees C produce males and those above 28 degrees C produce females (Vogt and Flores-Villela 1992).

Emydidae

This family comprises the river, pond, and box turtles. The family has approximately 100 species that are distributed on all major continents except Australia and Antarctica. However, nearly all species are found north of the equator. These animals like to bask on sunny days, often in large groups. They can be found in swift-flowing waters as well as slow-moving waters. Members of the family possess a complete set of plastral scutes (12 large, epidermal plates located on the bottom of the shell) and paddle-shaped hind feet. In the lowland wet forests of Costa Rica, these turtles are found in the slow-moving portions of rivers, alluvial plains, marshes, and estuaries. La Selva has two species of emydid turtles. One additional species is known from elsewhere in the Caribbean lowland wet forests of Costa Rica.

Key to the Emydidae at La Selva

1a Carapace domed, without longitudinal, middorsal ridge (fig. 11a); plastron dark brown with wide, median, yellow stripe or blotch (well beyond midventral suture); toes webbed. . . . . *Rhinoclemmys funerea*

1b Carapace unusually flat, with longitudinal, middorsal ridge (fig. 11b); plastron dark brown with no median, light (yellowish) stripe (at most, yellow associated with midventral suture); toes not webbed . . . . *Rhinoclemmys annulata*

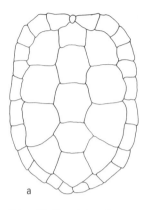
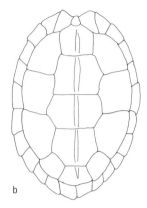

Figure 11. Dorsal view of carapace: (a) *Rhinoclemmys funerea* (no mid-dorsal ridge); (b) *Rhinoclemmys annulata* (with middorsal ridge).

BROWN WOOD TURTLE   *Rhinoclemmys annulata*
Pls. 82, 83

OTHER COMMON NAMES: Tortuga Café de Tierra, Galapago

SIZE: We have no data for specimens from La Selva, but elsewhere this species reaches sizes of up to 200 mm CL (Acuna 1993).

DISTRIBUTION: Caribbean drainages from Honduras through northern Colombia; Pacific drainages of Colombia and Ecuador.

IDENTIFYING FEATURES: The carapace of this species is extremely flat, possesses a low keel down the center, and is rugose because the scutes are thickened. In young individuals, the head is olive brown with a series of yellow lines and bars that extend onto the neck region. In older individuals, the colors become darkened and the lines disappear. The carapace is tan to dark brown, and the plastron is uniform dark brown. This species lacks webbing between the toes. The Brown Wood Turtle is similar in appearance to the Black Wood Turtle *(R. funerea)* but differs from that species by having a flatter shell, possessing little or no webbing on the feet, and lacking a yellow stripe down the center of the plastron.

SEXUAL DIMORPHISM: Like in many terrestrial turtles, the plastron of the male is concave compared with the flat plastron of the female. Also, the male can be distinguished from the female by the presence of a tail that, when stretched, has a cloacal opening located beyond the level of the carapace; the cloacal opening of the female extends only to the edge of the carapace.

HABITAT: This is an unusual turtle because it wanders the forest floor during the day rather than being found in or near water.

DIET: This species is largely herbivorous, consuming vines, shrubs, and ferns (Acuna 1993).

REPRODUCTION: Little is known of mating behaviors or nests in this species. However, the male has been observed to secrete saliva that is deposited on the head of the female and is thought to alter her behavior during copulation (Acuna 1993). Nesting occurs in the wet season (July to October), when a clutch of one to two eggs is placed in a shallow depression in the ground and/or covered with leaves (Acuna 1993).

REMARKS: This species often has ticks of the genus *Amblyomma* attached to the shell and the skin of the arms and legs.

BLACK WOOD TURTLE   *Rhinoclemmys funerea*
Pls. 84–86

OTHER COMMON NAMES: Tortuga Negra, Jicotea

SIZE: Largest individual 360 mm CL (n = 1).

DISTRIBUTION: Caribbean drainages of extreme southern Honduras to central Panama.

IDENTIFYING FEATURES: The head of younger individuals is dark olive green with yellow stripes on the neck and lateral surfaces of the face; these stripes are bordered by darker olive green. Older individuals have skin that is nearly uniformly black. The feet have well-developed webbing between the digits. The carapace is domed and dark brown to black. The carapace is often covered with algal growth, giving it a dull appearance. The plastron is dark brown with a yellow stripe in the center. This species can only be confused with the Brown Wood Turtle *(R. annulata)*, from which it differs by having a domed carapace, a yellow stripe down the center of the plastron, and webbing between the digits.

SEXUAL DIMORPHISM: The male can be distinguished from the female by the presence of a tail that, when stretched, has a cloacal opening located beyond the level of the carapace. The cloacal opening of the female extends only to the edge of the carapace.

Additionally, the plastron is concave in the adult male and flat in the adult female.

HABITAT: This is the common river turtle of La Selva. Adults are seen most easily during periods of low water when they bask on logs exposed below the suspension bridge over the Rio Puerto Viejo. Smaller individuals occur in the major creeks, e.g., the Sura at the Arboretum. In recent years the American Crocodile *(Crocodylus acutus)* has returned to the Rio Puerto Viejo, and the adult Black Wood Turtle has moved to smaller streams not occupied by the predator.

DIET: Although it is a common species, virtually nothing is known of its diet at La Selva. At other sites, it is herbivorous and forages terrestrially along river edges, often at night (Ernst and Barbour 1989).

REPRODUCTION: Courtship and copulation are known to take place in water, and the nesting season is from March to August, when one to six (usually three) eggs are deposited on the ground and covered with leaves (Acuna 1993).

REMARKS: This species often has ticks of the genus *Amblyomma* attached to the skin or between the platelike scales (scutes) that comprise the shell. A second species of emydid turtle, the Ornate Slider *(Trachemys ornata),* is characteristic of the rivers of the Caribbean lowlands of Costa Rica. It is surprising that specimens of this species have not been collected at La Selva. The Ornate Slider is likely to key to the Black Wood Turtle in our key, but does not conform to the species description of the latter species. See the species account for the Ornate Slider under "Additional Species" for any emydid that does not conform either to our key or to the species description for the Black Wood Turtle.

SQUAMATA

Squamates are reptiles in which males have paired copulatory structures called hemipenes and all individuals have kinetic skulls—skulls in which the bones of the upper and lower jaws move during feeding. This arrangement appears to be particularly successful: approximately 6,900 living species belong to this radiation. Two subgroups of squamates traditionally are recognized, one for lizards and one for snakes. However, several independent lineages of lizards have evolved snakelike features, one of them being the group to which snakes belong. Therefore, in evolutionary terms, snakes are lizards. For that reason, the traditional dichotomy within squamates is a false one. However, the convention of using the term snakes for limbless squamates with unique eyes and ears and using the term lizard for all other squamates is difficult to circumvent. Therefore, we use the terms lizard and snake to refer to lowland wet forest squamates with four legs and without legs, respectively.

Lizards and snakes have internal fertilization. A male everts one of his two hemipenes and inserts it into the cloaca of a female. Each hemipene is a thin-walled, saclike structure, the opening of which is attached to one side of the cloacal opening. Eversion of a hemipene involves using fluid pressure to force the structure out of the cloaca, turning it inside out in the process. The hemipene enters the cloaca of a female as it exits the opening of the male. When copulation is complete, contraction of a retractor muscle in the male returns the hemipene to its original position inside the base of the male's tail. Males actively court females with visual, tactile, and/or chemical cues, and females use these cues to select among potential mates. Courting behaviors are often performed in exposed places and can be easy to observe. Females allow males to mount them, a process that involves the male biting the female in the shoulder region and wrapping his tail around hers or, in limbless forms, involves the male wrapping his body around the female. Once inseminated, females of some species lay shelled eggs in a terrestrial or arboreal nest and may attend the nest while the eggs develop. Such eggs typically take 40 to 60 days to hatch. Offspring use a specialized egg tooth at the tip of the upper jaw to slice through the egg's shell. In other squamates, the eggs implant on the uterine wall and develop inside the female's reproductive tract. Some squamates with internal development have a thinner egg membrane, a feature that improves gas transfer, or a placenta, a feature that allows transfer of nutrients from the female parent to the offspring.

Some species of squamates are abundant and, therefore, are important prey items in food webs. Some species of snakes and birds specialize on squamates as prey. This is especially true of some birds of prey that feed nestlings almost exclusively squamates.

Three radiations of snakes have evolved salivary glands that are modified to deliver toxic substances to kill prey. These venoms can also be used to ward off potential predators. Viperid snakes have long fangs that fold into the front of the jaw when the mouth is closed. These snakes deliver venom that has a wide variety of toxic actions, including altering the ability of nervous tissue to transmit impulses, altering the ability of blood to coagulate, and digesting the connective tissue that holds cells together. Elapids have short, immobile fangs at the front of the upper jaw that deliver venoms that principally affect the nervous system. Finally, some colubrids (members of the xenodontine radiation) have enlarged teeth at the back of the upper jaw that deliver a variety of salivary chemicals with toxic effects of largely unstudied action.

All large squamates are capable of delivering painful bites that may result in deep puncture wounds or deep tissue damage. Such animals should be treated with respect because handling them involves a palpable risk. Small, nonvenomous squamates generally may be handled without risk to the handler. Some species of snakes and nearly all lizards have tails that break easily during handling. Additionally, the skin of some lizards is extremely delicate and may tear when handled.

Lizards are best captured by hand or noose. Simple nooses involve a thin string (e.g., dental floss) tied at one end to a thin stick and tied at the other end to create a loop that can be slipped over the head of a resting lizard and pulled tight around the lizard's neck. Snakes are best captured with the aid of a snake hook or tongs. However, small nonvenomous snakes can be pinned to the ground with a hand or stepped on gently until the head can be pinned with a stick. Such animals should be grabbed firmly behind the head and supported at mid body with the other hand.

Despite what you may see in charismatic nature shows, snakes should not be picked up and suspended by the tail. This procedure can damage the snake's backbone. Squamates can be kept for short periods of time in moistened cloth bags.

Key to the Families of Squamates at La Selva

1a   Four well-developed limbs 2
1b   No limbs.. 9
    2a   Dorsal scales shiny (smooth with half-moon-shaped posterior border or rectangular with weak keels; fig. 12a, b) 3
    2b   Dorsal scales dull (may be keeled, beaded, or velvety; fig. 12c, d) 4

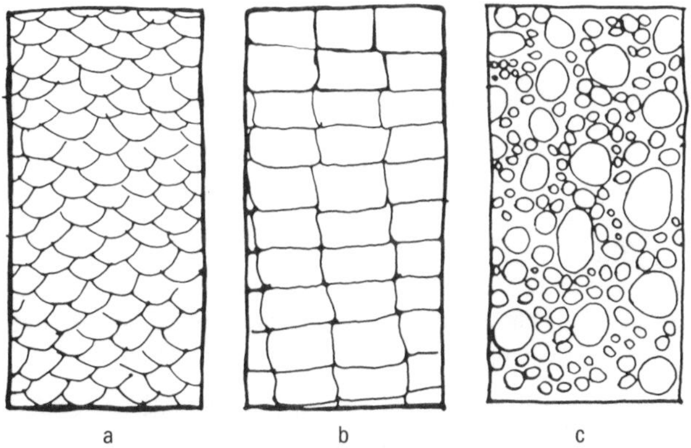

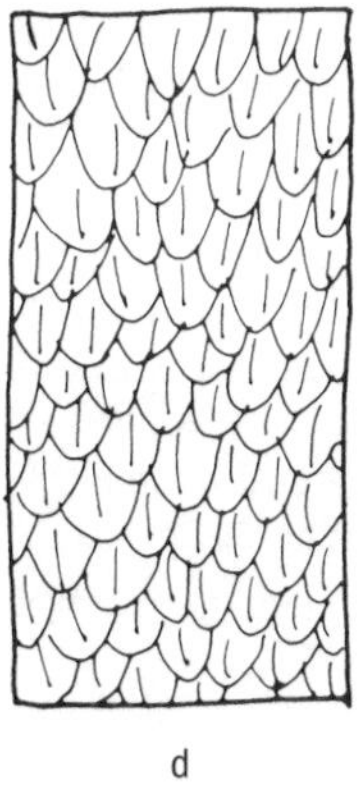

Figure 12. Representative samples of squamate skin: (a) smooth, cycloid scales; (b) rectangular scales; (c) beaded scales; (d) keeled scales.

3a Scales smooth and cycloid; no longitudinal fold of skin along side of body.............................Scincidae

3b Scales rectangular and keeled; a longitudinal fold of skin along side of body ....Anguidae

 4a Some dorsal scales beaded (rounded and much larger than matrix of surrounding scales; fig. 12c); body black to rusty tan with series of large, light spots .....
  ....Xantusiidae

 4b No beaded dorsal scales (scales more or less uniform, or if enlarged scales then restricted to middorsum); body not black to tan with light spots 5

5a Middorsal portion of head and neck with a series of spines or permanently erect crests in male and female 6

5b No spines or permanently erect crests on head and neck (although the male of some species erects low, temporary projections from the back of the head along the middorsum)
7

 6a Middorsal spines from back of head to tail...........
  ....Iguanidae

 6b Middorsal crests on back of head and neck ....
  ....Corytophanidae

7a Ventral scales enlarged and rectangular (fig. 12b); tail blue in the juvenile (faded in the adult) ....Teiidae

7b Ventral scales small and not rectangular; tail not blue 8

 8a Tail short (about equal to snout-to-vent length) and round or dorsoventrally compressed; no eyelid ......
  ....Gekkonidae

 8b Tail long (much longer than snout-to-vent length) and slightly laterally compressed; eyelids present.....
  ....Polychrotidae

9a A loreal pit between eye and nostril (fig. 13a); enlarged, mobile fangs at front of mouth; head triangular..............
 ....Viperidae

9b No loreal pit (fig. 13b); no fangs, or if fangs then immobile; head not noticeably triangular (unless harassed, in which case some species flare quadrate and squamosal bones to create a triangular head shape)........................10

 10a Short, immobile fangs at front of mouth; small black eye barely or not at all visible in black head cap.......
  ....Elapidae

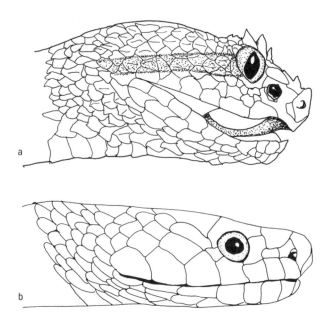

Figure 13. Lateral view of head: (a) *Bothriechis schlegeli* (showing loreal pit and enlarged supraocular scales); (b) *Clelia clelia* (no loreal pit, no enlarged supraocular scales).

10b No fangs at front of mouth; eyes usually large and not hidden in black coloring of head 11

11a Scales on top of head between and behind eyes small, similar in size to dorsal scales on body; ventral scutes narrow, barely covering width of belly 12

11b Scales on top of head between eyes platelike and noticeably larger than dorsal scales of body; ventral scutes wide, covering entire width of body . . . . Colubridae

12a Scales between nares larger than scales covering rest of head . . . . Ungaliophiidae

12b Scales between nares small, same size as scales covering rest of head . . . . Boidae

Corytophanidae

This family contains nine species of midsized to large lizards that have a single or bi-lobed, enlarged, fanlike projection from the back of the head to the middorsum. In the genus *Basiliscus*, this projection is better developed in males than in females. In the other two genera, *Corytophanes* and *Laemanctus*, the projections are equally well developed in males and females. Corytophanids are restricted in distribution to Central America and northern South America. Members of this family tend to be found in vegetation along or near rivers and streams. All lay eggs in ground nests. La Selva has three species of corytophanids, and they represent the fauna known from the rest of the Caribbean lowland wet forests of Costa Rica as well.

Key to the Corytophanidae at La Selva

1a Body green with double-lobed crest (anterior lobe small)
. . . . *Basiliscus plumifrons*
1b Body brown with single-lobed crest 2
  2a Border of crest with enlarged, serrate scales. . . . .
  . . . . *Corytophanes cristatus*
  2b Border of crest smooth . . . . *Basiliscus vittatus*

GREEN BASILISK
Pl. 87

OTHER COMMON NAMES: Jesus Christo, Chisbala, Garrobo, Jesus Christ lizard
SIZE: Largest male 155 mm SVL (n = 2); largest female 77 mm SVL (n = 1). This species is known to reach sizes of up to ca. 200 mm SVL.

Basiliscus plumifrons

DISTRIBUTION: Caribbean slopes from Honduras to Panama.
IDENTIFYING FEATURES: The adult Green Basilisk is easy to identify because of the presence of large crests (very large in the male), two on the back of the head and neck and one along the length of the back and base of the tail. In the adult male the body is leaf green with a series of light blue spots. The juvenile and the adult female are more drab in coloration, usually olive brown with a series of lateral, black bars and blue green spots. The adult has a

bright yellow iris. All individuals have a uniform yellowish green venter that is very bright in the juvenile. The juvenile can be confused with the juvenile Green Iguana *(Iguana iguana)*, but the latter species lacks blue spots on the sides of the body and has small, spikelike, middorsal scales (not present in the Green Basilisk).

SEXUAL DIMORPHISM: The male is much larger than the female. The adult male has a much larger head crest (larger than the head in the male; smaller than the head in the female), has a large, fanlike projection along the middorsum (a uniform low ridge in the female), and is brighter green than the female.

HABITAT: The Green Basilisk can be found along any of La Selva's rivers and major streams. It is commonly seen in the vegetation along the Rio Puerto Viejo at the River Station. This animal is difficult to approach during the day but can be captured while sleeping on its night roost in a small tree or shrub along a stream or river. The juvenile is often found in the patches of wild banana (*Heliconia* spp.) in the laboratory clearing.

DIET: The diet of this lizard is catholic (Hirth 1962), including invertebrates, vertebrates (even bats!), and vegetation (mostly fruits and flowers).

REPRODUCTION: The female digs a hole within which eggs are placed and covered. Three eggs collected by H. W. Greene on February 15, 1983, hatched on May 4 (two eggs) and May 5 (one egg); hatchlings were 43 to 47 mm SVL at birth. Because juveniles are found so frequently in the laboratory clearing, we suspect that females may frequently nest here.

REMARKS: This animal gets one of its English common names, the Jesus Christ Lizard, from its ability to locomote bipedally across water. The fourth toe of each hind foot is elongate and broadened, with lamellae on the bottom of the toe. This feature and the wide spread of the toes allow the animal's hind feet to deflect the water's surface downward, creating an air pocket. Sufficient thrust is generated from each hind leg to maintain the body above water and to withdraw the foot before the air pocket collapses. This process is easier to perform in small juvenile lizards than large adults (Glasheen and McMahon 1996).

STRIPED BASILISK
Pls. 88, 89

Basiliscus vittatus

OTHER COMMON NAMES: Jesus Christo, Cherepo, Basilisk, Jesus Christ lizard, Brown Basilisk

SIZE: No specimen from La Selva has been measured. Elsewhere in Costa Rica this species reaches sizes of ca. 140 mm SVL.

DISTRIBUTION: Caribbean slopes from Tamaulipas, Mexico, to northern Colombia; Pacific slopes from Jalisco, Mexico, to Nicaragua.

IDENTIFYING FEATURES: This lizard cannot be confused with any other species. A single, large, fanlike crest runs along the back of the head and neck of the male (reduced in size in the juvenile and female). The dorsum is uniform tan to dark chocolate brown, except for a distinctive light yellow stripe that starts at the tip of the snout, includes the supralabials, and continues along the side of the head, neck, and body to the level of the armpit (axilla).

SEXUAL DIMORPHISM: The male is larger than the female (Lee 1996). Also, the male has a large crest from the top of the head to the back of the neck (as large as the size of the head); the female has a weak ridge in place of the crest.

HABITAT: This species recently has been observed in disturbed areas near the cafeteria and on the La Flaminea annex. To date it has not been found in the forested habitat to the north of the Rio Puerto Viejo or to the east of the Rio Sarapiquí. It is typically associated with open habitats along the Caribbean coast and apparently moves inland along areas disturbed by humans. This species is usually observed basking along pools, creeks, and rivers or in open areas associated with houses and towns. At night it can be found perched on shrubby vegetation.

DIET: The diet consists of a variety of insects (mostly ants in the adult) and vegetation, including fruits (Hirth 1962).

REPRODUCTION: Up to 12 eggs are laid in a ground nest late in the dry season (Alvarez del Toro 1960). The female may produce more than one clutch each season (Leenders 2001).

REMARKS: When disturbed this lizard escapes with bipedal locomotion on land or across the surface of water, hence one of the English common names, Jesus Christ Lizard. This species is rare at La Selva but may be invading disturbed areas. It appeared on the early species lists for the station (see Scott et al. 1983) but was not seen for a long period of time after that (Guyer 1994b). Because the body shape and coloring of this species are distinctive, it is not likely to have been confused with any other species.

CASQUE-HEADED LIZARD   *Corytophanes cristatus*
Pl. 90

OTHER COMMON NAMES: Perro Zonpopo, Smoothhead Helmeted Basilisk

SIZE: Largest male 98 mm SVL (n = 2); largest female 116 mm SVL (n = 1).

DISTRIBUTION: Caribbean slopes from Veracruz, Mexico, to northern Colombia; Pacific slopes from the Osa Peninsula, Costa Rica, to northwestern Colombia.

IDENTIFYING FEATURES: Both sexes have a single, large crest from the back of the head to the anterior portion of the trunk. The posterior margin of the crest is composed of pointed (serrate) scales that are confluent with a series of enlarged middorsal scales that continue the length of the animal. A similar set of serrate scales is found along the midventral line of the chin. The dorsum is mottled with dark and light brown, yellow, red, and green. The body coloring and shape make this lizard similar in appearance to a dead leaf. The iris is brick red in color. This lizard is capable of changing color rapidly. This species is unlikely to be confused with any other at La Selva.

SEXUAL DIMORPHISM: No known external features distinguish the two sexes.

HABITAT: This lizard is diurnal and arboreal, perching low on understory shrubs and small trees. We have observed it in both primary and secondary forests. Because this species is so cryptic, it is rarely observed, although it may be common. It tends not to move even when approached, so it is most often found by observers who are suddenly surprised to realize that the "dead leaf" they had been ignoring is actually a lizard. This animal can also be found at night by shining a light on understory vegetation, where the lizard sleeps on thin, upright stems. The long hind legs, when folded, stick out from the perch and often reflect light more brightly than the vegetation.

DIET: This odd looking creature is a classic sit-and-wait predator. It remains motionless for days, waiting for large insects (often caterpillars, beetles, and tettegoniids; Andrews 1979) and occasionally small lizards (Leenders 2001) to come within reach.

REPRODUCTION: The Casque-headed Lizard lays five to six eggs, each approximately 20 mm in length, in a nest dug in the ground; eggs may take 6 months to hatch (Leenders 2001). Some suggest that the casque head of the female is used to dig the nest (Bock 1987). One hatchling measured from La Selva was 43 mm SVL.

Iguanidae

Iguanids are large, arboreal, herbivorous lizards possessing a series of spines down the middle of the back from the neck to the base of the tail. This family contains 34 species distributed from the desert southwest of North America, through Central America, to Amazonian South America. A disjunct member of the family is also found on the Fiji Islands. These lizards are social creatures that often are organized into harems (several females within the territory of a dominant male). Males display frequently to one another and to females. These displays separate males into distinct territories within which females cluster. Females lay eggs in ground nests. These lizards can be abundant and serve as important food sources for top predators, such as jaguars and humans. At La Selva only a single species of this family is present, and it represents the fauna known from the rest of the Caribbean lowland wet forests of Costa Rica as well.

GREEN IGUANA *Iguana iguana*
Pls. 91, 92

OTHER COMMON NAMES: Gallena de Palo,
Iguana Verde

SIZE: We have no measurements from La Selva; elsewhere the male reaches ca. 400 mm SVL and the female reaches ca. 300 mm SVL (Lee 1996).

DISTRIBUTION: Caribbean slopes from Veracruz, Mexico, to southern Brazil and Paraguay; Pacific slopes from Sinaloa, Mexico, to Ecuador.

IDENTIFYING FEATURES: The Green Iguana is unmistakable, given its large size (the largest lizard in Costa Rica), the presence of a ridge of spines from the neck to the base of the tail (long and spikelike in the male), and the presence of a greatly enlarged scale at the back of the lower jaw below the tympanum. The adult male usually has an orangish gray head and a gray green body marked by a series of vertical, dark bars along the sides. The tail coloration consists of a series of light grayish green and dark gray to black rings; these can become faded in larger adults. The female and juvenile are bright green with a series of dark, vertical bars along the sides and have banded tails. This color pattern is similar to the Green Basilisk *(Basiliscus plumifrons),* but the Green Iguana can be distinguished from that species by the presence of small, spikelike, middorsal scales (not present in the Green Basilisk).

SEXUAL DIMORPHISM: The male is larger than the female and becomes orangish in color (especially on the spines) during the mating season (dry season). Additionally, the male has a dewlap, a flap of skin below the chin, that is presented to rivals or potential mates during bobbing displays and is often conspicuous when the male is perched in its territory.

HABITAT: The easiest place to observe the Green Iguana is from the suspension bridge over the Rio Puerto Viejo, where several territorial males have created matted perches on tree limbs (often very high up) overhanging the river. A careful survey of the trees usually allows observation of many females, but these are cryptic and difficult to spot. Juveniles often are found in great abundance in the laboratory clearing, which suggests that nesting takes place away from water. The species is strictly diurnal in activity.

DIET: The Green Iguana is largely herbivorous (Iverson 1982; Rand et al. 1990) but the juvenile augments its diet with invertebrates (Hirth 1963). Foraging takes place during short (one to five minute) bouts, usually in a single tree that may be carefully selected from among those in the animal's home range; leaves, flowers, and/or fruit may be consumed (Rand et al. 1990). Bacteria that aid in the digestion of cellulose occur in the hind gut and are passed to juveniles via coprophagy (Troyer 1982).

REPRODUCTION: The male establishes a territory during the dry season by giving head-bobbing displays from exposed positions high in the canopy trees along major rivers and streams. These displays are also used to court females. Females aggregate within the territories of the largest males and establish a dominance hierarchy within the aggregation. Mating takes place in the trees, and a male copulates only once on any one day (Rodda 1992). The fertilized female then creates a ground nest in which 20 to 70 eggs are placed (Alvarez del Toro 1960; Carr 1954; Hallinan 1920; Hirth 1963; Swanson 1950). The nest consists of a hole up to 1 m deep—so deep that the female may dig a separate exit (Hirth 1963). The nesting sites of the Green Iguana at La Selva are unstudied but probably include sand bars and the clearings along major rivers. At other sites female Green Iguanas migrate synchronously to open areas to dig nests (Bock and Rand 1989; Henderson 1974; Hirth 1963; Rand 1972). Incubation takes approximately three months. Hatchlings emerge synchronously, depart nest sites in groups, and spend their early development in open areas away from the sites occupied by adults (Burghardt et al. 1977; Henderson 1974).

REMARKS: A large adult often plummets from its perch when approached by potential predators (e.g., humans floating down rivers). An individual displaying this behavior often careens off branches during its descent and can land with a resounding splash because its lungs are inflated when it hits the water. The animal may then swim rapidly to shore or dive. We once observed a single adult male that was emaciated, apparently from disease. Some local residents of the Sarapiquí region describe periodic episodes of mass mortality in local Green Iguana populations.

Polychrotidae

This is the family to which the anoles and their allies belong, a species-rich radiation containing at least 450 species. The family is found from the southeastern United States, through Mexico and Central America, to near the southern tip of South America; nearly all of the Caribbean islands are occupied by members of this family. The family is distinguished by the presence of a fan-like flap of skin on the chin, called a dewlap, that is attached to the hyoid bones. The dewlap can be extended and retracted, and is usually colorful in males and drab when present in females. This structure is used by males in territorial and courtship displays. The family Polychrotidae is also characterized by small, velvety scales on the dorsum and venter. However, unlike the family Gekkonidae (geckos), which includes species with similar velvety skin, the polychrotids have a tail that is long, thin, and slightly compressed laterally. Many species of polychrotids co-occur in the same habitat. In the lowland wet forests of Costa Rica, these syntopic forms do not conform to the distinctive ecomorphs of island polychrotids (Beuttell and Losos 1999; Williams 1983). All members of the family lay eggs, and most are highly arboreal. With nine species, this is the most species-rich family of lizards at La Selva. One additional species is known from elsewhere in the Caribbean lowland wet forests of Costa Rica, but it was introduced to the region by humans. The older literature refers to all of these lizards, except the Neotropical Chameleon *(Polychrus gutturosus)*, as belonging to the genus *Anolis*. Because the Central American anoles have a unique set of caudal vertebrae (Etheridge 1967), we follow the current trend of separating these forms into their own genus, *Norops* (Guyer and Savage 1986, 1992).

Key to the Polychrotidae at La Selva

1a Nostrils on a rounded swelling (best viewed laterally) . . . . .
. . . . *Norops capito*

1b Nostrils not on a distinctive, rounded swelling. 2

    2a Basic body color faint to distinct lime or green when first captured 3

    2b Basic body color brown, tan, gray, or dull olive 5

3a White patch from back of eye to corner of mouth; sides with light, diagonal bars bordered by dark gray; dewlap of male gray with enlarged, yellow green scales . . . .
. . . . *Polychrus gutturosus*

3b No white patch from back of eye to corner of mouth; dewlap of male variously colored but not gray with enlarged, yellow green scales 4

4a Color pattern lichenose (mottled green, yellow, brown, and/or black); dewlap of male orange; dewlap of female small and white with an orange wash . . . . *Norops carpenteri*

4b Color pattern more or less uniform green (color may change rapidly in hand to include brown); dewlap of male light blue to purple with a wide, orange border and with rows of enlarged white scales; dewlap of female light blue with enlarged white scales. *Norops biporcatus*

    5a A lateral, light stripe from tip of snout to near groin . .
. . . . *Norops oxylophus*

    5b No lateral, light stripe from tip of snout to groin 6

6a Middorsal scales distinctly larger than other scale rows, not grading from larger to smaller (fig. 14); dewlap of male red to orange with a yellow border . . . . *Norops humilis*

6b Middorsal scales about the same size as other scale rows or grading gradually from slightly larger to smaller; dewlap of male not red bordered by yellow 7

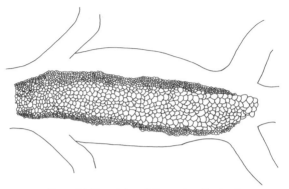

Figure 14. Dorsal view of *Norops humilis* showing relative size of enlarged middorsal scale rows.

7a Upper lip white; body extremely thin; dewlap of male white with a faint orange or yellow spot in center. . . . . .
. . . . *Norops limifrons*

7b Upper lip not lighter than rest of head; body not noticeably thin; dewlap of male not white with an orange or yellow center. 8

8a Hind legs noticeably short, longest hind toe reaching no further than shoulder when leg adpressed against side (fig. 15a); dewlap of male magenta or purple and large (as large as or larger in surface area than head); lining of mouth cavity black . . . . *Norops pentaprion*

8b Hind legs not noticeably short, longest toe reaching further than shoulder when leg adpressed against side (fig. 15b); dewlap red with black scales in male, white to light yellow in female, and small in both sexes (smaller in surface area than head); lining of mouth cavity not black . . . .
. . . . *Norops lemurinus*

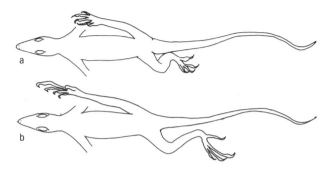

Figure 15. Stylized dorsal view of genus *Norops:* (a) short hind limbs; (b) long hind limbs.

GREEN TREE ANOLE   *Norops biporcatus*

Pls. 93, 94

OTHER COMMON NAMES: Neotropical Green Anole

SIZE: Largest male 94 mm SVL (n = 2); largest female 95 mm SVL (n = 2). Leenders (2001) indicates that this species can reach sizes of 106 mm.

DISTRIBUTION: Caribbean slopes from Chiapas, Mexico, to northern South America; Pacific slopes from Nicaragua to Ecuador.

IDENTIFYING FEATURES: This lizard is large for an anole. The body coloring is distinctive in that an animal freshly captured in the field invariably is leaf green. This species has a ring of yellow scales around each eye and often a series of light blue spots along the sides of the body. The body coloring can (and usually does) change rapidly to mottled green interspersed with light and dark brown (much to the chagrin of nature photographers). The tail is distinctly banded light and dark in the juvenile; this banding is faded in the adult such that the tail is more or less uniform in color. The iris is dark reddish brown, and the hind limbs are noticeably short, with the longest toe reaching no farther than the shoulder when the leg is adpressed against the animal's side. The dewlap of the male is light purple with a wide, orange border; the enlarged scales of the dewlap are white. In the female, the dewlap is small and light blue with enlarged, white scales. The only other uniform green anole at La Selva is Carpenter's Anole *(N. carpenteri)*, which is smaller in size, thinner in body shape, and lighter (lime) green in body color, and which has an orange dewlap in the male.

SEXUAL DIMORPHISM: The sexes can be distinguished by the color of the dewlap (see above). There is no noticeable difference in size between the male and female in populations in Costa Rica (Corn 1981), but the male is slightly larger than the female at Los Tuxtlas (Villareal-Benitez 1997).

HABITAT: This species inhabits the large branches of canopy trees (Villareal-Benitez 1997); most individuals encountered at La Selva are on or near newly fallen trees. This lizard is diurnal in its activity pattern.

DIET: Invertebrates, especially coleopterans and arachnids, are the primary diet items, although this lizard occasionally eats small fruits (Villareal-Benitez 1997).

REPRODUCTION: Like all anoles, the Green Tree Anole lays a single egg per clutch, probably in epiphytes or leaf litter. Villareal-Benitez (1997) suggests that reproduction occurs primarily in the dry season at Los Tuxtlas.

REMARKS: This species emits a squeaklike vocalization when handled (Myers 1971).

PUG-NOSED ANOLE   *Norops capito*
Pls. 95, 96

OTHER COMMON NAMES: Bighead Anole
SIZE: Largest male 88 mm SVL (n = 15);
largest female 93 mm SVL (n = 12).
DISTRIBUTION: Caribbean slopes from Tabasco,
Mexico, to Panama; Pacific slopes from the Osa
Peninsula, Costa Rica, to Panama.

IDENTIFYING FEATURES: This species is characterized by a distinctive, short, blocklike head. A small ridge of bone above each eye and projecting laterally (easiest to see when viewed head-on) is unique to this species. Additionally, the external nares are located on a swollen projection at the tip of the snout; this feature makes the forehead look deeply scalloped when viewed from the side. Often there is a distinctive (in contrast to the rest of the head), dark, interorbital bar. The male has a small, burnt orange—and the female a small, olive green—dewlap. The enlarged scales of the dewlap are white. The body coloring is extremely variable but usually has a lichenlike or mosslike pattern. Most individuals are green mixed with shades of gray, tan, and/or dark brown, but a few individuals lack green; this complex color pattern can change rapidly in the hand. Sometimes the female has a wide, tan, mid-dorsal stripe, bordered by light and dark stripes. The iris is brick red to orange in both sexes, and the hind legs are long, with the longest toe reaching beyond the eye when the leg is adpressed against the side of the body. The head shape of the Pug-nosed Anole distinguishes this species from all other anoles at La Selva.

SEXUAL DIMORPHISM: The sexes can be distinguished by the color of the dewlap (see above). Additionally, the female is slightly longer in body length than the male (Corn 1981). In the adult male, the base of the tail is noticeably swollen to house the hemipenes, whereas the base of the tail tapers evenly in the female.

HABITAT: The hind legs of this species are long, suggesting great leaping ability, but little is known of the species's habitat utilization. The species occurs in primary and secondary forest and is the third most commonly encountered anole in abandoned cacao *(Theobroma cacao)* plantations (behind the Ground Anole *[N. humilis]* and the Slender Anole *[N. limifrons]*). Typically, this species is diurnal, is seen on or near tree trunks, and invariably attempts to escape by running upward. When on the ground, it hops rather than runs, and we have observed it jumping from frond to frond in tall understory palms.

DIET: This lizard eats formicids and dipterans (Lieberman 1986) and occasionally other anoles (Andrews 1983).

REPRODUCTION: Because the juvenile is often encountered in the leaf litter or at the base of trees, we suspect that eggs are laid on the ground under the leaf litter. Gravid females have been collected during all seasons (Corn 1981), but insufficient data are available to describe patterns of reproduction.

CARPENTER'S ANOLE   *Norops carpenteri*

Pls. 97, 98

SIZE: Largest male 41 mm SVL (n = 6); largest female 43 mm SVL (n = 8).

DISTRIBUTION: Caribbean slopes from Nicaragua to western Panama.

IDENTIFYING FEATURES: The ground color is a light lime green mottled with white and tan flecks. Often the female has a thin, indistinct, bronze, middorsal stripe. The tail is banded with light gray and darker grayish green markings. This species is quite slender in body shape and has unusually short hind legs, such that the fourth toe of the hind leg does not extend past the tympanum when the leg is adpressed against the side of the body. The skin surrounding the eye is yellow. The dewlap of the male is burnt orange; the female has a small, white dewlap that has a faint orange wash. Because it is also slender and has a banded tail, the Slender Anole *(N. limifrons)* is the most similar species to Carpenter's Anole at La Selva. However, the former species is tan in body color, has long hind limbs (with the longest toe reaching well beyond the eye when the leg is adpressed against the side of the body), and possesses a distinct, light stripe along the upper lip to the shoulder. The only other uniform green anole at La Selva is the Green Tree Anole *(N. biporcatus)*, which is larger in size (as an adult), is a more vivid, leaf green in body color, and has a purple dewlap with an orange margin.

SEXUAL DIMORPHISM: The sexes can be distinguished by the coloring of the dewlap (see above). The body of the adult female is slightly longer than that of the adult male (Corn 1981).

HABITAT: This species is apparently uncommon throughout its range, for it was not described until 1971 (Echelle et al. 1971). Echelle et al. (1971) collected specimens from the vegetation and moss-covered rocks along riverbanks. At La Selva, we have observed this lizard in a variety of habitats (primary forest, cacao *[Theobroma cacao]* plantations with pejiballe *[Bactris gasipaes]* and laurel *[Cordia alliodora],* and the Arboretum) and on diverse substrates (trunks, twigs, and leaves). It is unusual among La Selva's anoles in that, when disturbed, it may freeze and/or drop to the ground. It has also been observed to escape by running up tree trunks or swimming (Echelle et al. 1971). The species is diurnal in activity pattern.

DIET: The diet of this species consists of small arthropods, dominated by homopterans, orthoperans, and araneids (Corn 1981).

REPRODUCTION: Gravid females have been observed throughout the year, but data on testes size in males suggest a late wet season diminution of reproduction (Corn 1981). Eggs presumably are deposited in the leaf litter or in small crevices on tree trunks.

GROUND ANOLE
PI. 99

SIZE: Largest male 36.5 mm SVL (n = 216); largest female 40 mm SVL (n = 222).

DISTRIBUTION: Caribbean slopes from Honduras to Panama; Pacific slopes of southern Costa Rica.

Norops humilis

IDENTIFYING FEATURES: This lizard is dark brown with (1) a series of one to four dark spots or V's in the sacral region (male or female), (2) a wide, dark gray stripe down the middle of the back (female only), or (3) a series of dark brown diamonds down the middle of the back (female only). In the male, the dewlap is bright red to orange with a yellow border; the female lacks a dewlap but may show a reddish wash in the throat region. The iris is dark brown with a copper border around the pupil. Two useful field characteristics allow easy identification of the Ground Anole. One is the presence of a series of enlarged scales down the middle of the back (a hand lens is needed to see this feature on smaller individuals). These scales are more than twice as large as the granular scales of the rest of the body. The second characteristic is the presence of a large, tubelike invagination in the armpit (axilla). The function of this structure is difficult to determine; it is often filled with mites and/or ticks. This species is most likely to be confused with the Lemur Anole *(N. lemurinus)*, but the latter species lacks the enlarged dorsal scales and armpit invagination of the Ground Anole.

SEXUAL DIMORPHISM: The sexes can be distinguished by the dewlap (see above). The female matures at a longer body length (30 mm SVL) than the male (25 mm SVL; Guyer 1986).

HABITAT: This is one of two common anoles at La Selva (the other being the Slender Anole *[N. limifrons]*). It is found on the leaf litter, low in vegetation, or on tree trunks, usually within 1 m of the ground. The Ground Anole is a deep-forest species (Talbot 1977) but can also be found in great abundance in the Arboretum and in some regions of the abandoned cacao *(Theobroma cacao)* plantation (more frequently in areas mixed with pejiballe *[Bactris gasipaes]* and laurel *[Cordia alliodora]*). This lizard is diurnal in activity pattern.

DIET: This species eats a variety of small to midsized arthropods, including formicids, adult and larval coleopterans, homopterans, acarines, centipedes, and isopods, but especially araneids and hemipterans (Lieberman 1986).

REPRODUCTION: Reproduction occurs throughout the year but peaks during the dry season, when a single-egg clutch may be produced as often as once per week and placed in the leaf litter (Guyer 1986, 1988b). Two eggs laid on July 29, 1982, hatched after 46 and 57 days. The hatchling is as small as 16 mm at birth.

LEMUR ANOLE
Norops lemurinus
Pls. 100–102

OTHER COMMON NAMES: Ghost Anole, Canopy Anole

SIZE: Largest male 51 mm SVL (n = 10); largest female 62 mm SVL (n = 15).

DISTRIBUTION: Caribbean slopes from Veracruz, Mexico, to Panama; Pacific slopes of Costa Rica and western Panama.

IDENTIFYING FEATURES: This species is difficult to distinguish from other anoles because of its variable color pattern and intermediate morphology. However, the body of captive animals is typically tawny with a yellow wash on the sides and venter. This species usually has a dark bar on the head between the eyes and a dark, U-shaped mark on the back of the head and neck. The dewlap of the male is deep red with a series of enlarged, black scales, and the female has a well-developed, white dewlap. The female usually has a series of triangular markings or a wide (light or dark) stripe down the middorsum. The tail of the adult is weakly banded with light and dark, and the hind limbs are of moderate length, with the longest toe reaching near the eye when the leg is adpressed against the animal's side. The iris is dark brown with a hint of red. This species is most similar to the Ground Anole *(N. humilis)* and the Slender Anole *(N. limifrons)*. The Ground Anole can be distinguished from the Lemur Anole by the presence of enlarged middorsal scales and a deep pocket in the axilla (armpit). The Slender Anole differs from the Lemur Anole in being thinner, having longer hind limbs (the longest toe reaches well beyond the eye when the leg is adpressed against the animal's side), and possessing a boldly banded tail (a distinguishing characteristic only when the tail has not been lost and replaced).

SEXUAL DIMORPHISM: The sexes can be distinguished by the color of the dewlap (see above). The female has a longer body on average than the male (Corn 1981).

HABITAT: The species frequents the trunks of large trees and is especially abundant (or easy to see) in the Arboretum. It invariably attempts to escape by running upward. We have also recorded this species from the abandoned cacao *(Theobroma cacao)* plantation, the pejiballe trees *(Bactris gasipaes)* in the lab clearing, and the plantings of laurel trees *(Cordia alliodora)* in the Huertos plots.

DIET: The diet is dominated by coleopterans, orthopterans, lepidopteran larvae, and araneids (Corn 1981).

REPRODUCTION: A reduction in the number of juveniles observed late in the wet season suggests a decline in reproduction during that time (Corn 1981).

REMARKS: For those at sites along the Caribbean coast of Costa Rica, any anole that keys to the Lemur Anole in our book should be compared with the description of the Puerto Rican Crested Anole *(Ctenonotus cristatellus)* (see under "Additional Species"). The Puerto Rican Crested Anole was introduced to the port of Limón and has since spread northward and southward along the coast of Costa Rica.

SLENDER ANOLE   *Norops limifrons*

Pls. 103, 104

OTHER COMMON NAMES: Cherepo

SIZE: Largest male 41 mm SVL (n = 102); largest female 43 mm SVL (n = 46).

DISTRIBUTION: Caribbean slopes from Honduras to Panama; Pacific slopes from Costa Rica to Panama.

IDENTIFYING FEATURES: The Slender Anole gets its common name from its extremely thin, seemingly delicate appearance. The hind legs are very long, the tail is long, and the trunk is thin. The body usually is grayish brown to light olive or tan. The female sometimes has a middorsal, light yellowish stripe bordered by dark brown or has a series of diamond-shaped, dark markings middorsally. The upper lip is often white, appearing as a distinct stripe, and the tail is often ringed with light and dark (this trait is not present in replacement tails). The venter is white or lemon yellow, and the iris is copper to orange. The male has a small, white dewlap with an orange (sometimes yellow) spot in the center. The female has no dewlap. This species is most similar in size and shape to Carpenter's Anole *(N. carpenteri),* but the latter is lime green in color, has much shorter hind legs (with the longest toe not reaching the eye when the leg is adpressed against the side of the body), and has an orange dewlap in the male. The Lemur Anole *(N. lemurinus)* also can be confused with the Slender Anole, but the former species has a thicker body, shorter hind legs (such that the longest toe does not extend beyond the eye when the leg is adpressed against the side of the body), and a red dewlap in the male.

SEXUAL DIMORPHISM: The two sexes can be distinguished by the dewlap (see above). The male can also be distinguished from the female by the presence of a pair of enlarged postanal scales, but a hand lens is necessary to see these structures. The adult female of this species is slightly longer than the adult male (Corn 1981).

HABITAT: This is the most commonly observed anole at La Selva. This animal leaps from perch to perch in shrubby areas. It is found in primary forest, as well as in second growth, including open areas such as the laboratory clearing (Talbot 1977). The Slender Anole leaps through the understory or runs up tree trunks to escape predators.

DIET: At La Selva, this lizard eats small invertebrates, especially adult and larval dipterans, coleopterans, and araneids, but also including blattids, orthopterans, hemipterans, acarines, and isopods (Lieberman 1986). Often prey are captured on the surfaces of leaves.

REPRODUCTION: Gravid females have been observed during all months of the year. However, at Rio Frio, juvenile abundance declines late in the wet season, suggesting reduced reproduction at that time (Corn 1981).

REMARKS: Male-female pairs often chase one another. This observation might indicate that pair bonds are formed (Talbot 1979). The population density of this species fluctuates dramatically among years on Barro Colorado Island (Andrews 1991).

STREAM ANOLE

Norops oxylophus

Pl. 105

SIZE: Largest male 79 mm SVL (n = 13); largest female 60 mm SVL (n = 14).

DISTRIBUTION: Caribbean slopes from Honduras to Panama; Pacific slopes of central Costa Rica.

IDENTIFYING FEATURES: This distinctive species is dark olive brown in ground color with a light yellow or cream stripe on each side starting along the upper lip and continuing posteriorly along the side of the body to the level of the groin. The venter of the chin is white, shading to light yellow or cream posteriorly. The belly is light yellow or cream, shading to peach laterally. The dewlap in the male is large and yellowish orange in color; the enlarged scales of the dewlap are white to cream in color. The female has a small dewlap. The tail is uniform olive brown, and the iris is reddish brown. The body color and lateral stripe distinguish this species from all other anoles at La Selva.

SEXUAL DIMORPHISM: The sexes can be distinguished by the dewlap (see above). Also, the male possesses a pair of enlarged postanal scales, located on the ventral surface of the tail posterior to, but near, the cloacal opening, that are visible with a hand lens. Additionally, the male may display nuchal and caudal crests during social interactions and may do so upon capture as well. Unlike most of La Selva's anoles, the adult male of this species is longer than the adult female (Corn 1981); in Nicaragua the female matures at 49 mm and the male at 53 mm SVL (Vitt et al. 1995).

HABITAT: This lizard can be found on moss-covered tree trunks, logs, and rocks along streams and rivers. It is common along the Quebrada Sura at the Arboretum and can be found in the Research swamp. When escaping, it may dive underwater and remain submerged for several minutes.

DIET: This lizard consumes a wide variety of arthropods, but coleopterans, lepidopteran larvae, dipterans, formicids, and araneids predominate (Corn 1981; Vitt et al. 1995).

REPRODUCTION: The female lays each egg under moss growing on rocks along streams, sometimes with several females sharing the same communal site (Campbell 1973). There is no evidence of seasonality of reproduction in this species (Corn 1981).

REMARKS: The specific epithet used here reflects the revision of the aquatic anoles by Williams (1984). Literature before Williams (1984) refers to this species as *N. lionotus*.

LICHEN ANOLE

Pls. 106, 107

Norops pentaprion

SIZE: Largest male 61 mm SVL (n = 4); largest female 80 mm SVL (n = 3).

DISTRIBUTION: Caribbean slopes from the Isthmus of Tehuantepec, Mexico, to eastern Panama; Pacific slopes from the Isthmus of Tehuantepec, Mexico, to northwestern Colombia.

IDENTIFYING FEATURES: This anole is relatively wide and flat and has a mottled gray and brown dorsum with a ringed (light tawny and dark brown) tail. The color pattern looks like lichens on tree trunks. The venter of the body, hind limbs, and anterior portion of the tail is tawny yellow. When the mouth is open, a black oral lining and light blue skin at the angle of the jaws become visible. The male has a large, magenta dewlap; the female has a dewlap, but it is smaller than that of the male. This species has short hind legs that are about the same length as its arms; the longest toe of the hind leg reaches no farther than the shoulder when the leg is adpressed against the side of the body. It also has a tail that is relatively short and may wrap around objects, giving the impression that it is prehensile. The skin in the axillary (armpit) region is flaplike, creating an airfoil that may stabilize the animal during long leaps from perch to perch.

SEXUAL DIMORPHISM: In the male, a low, middorsal crest may be raised from the back of the head to near the tip of the tail. The sexes can also be distinguished by the color of the dewlap (see above). Corn (1981) documented that the male is larger than the female at Rio Frio.

HABITAT: This is the only species of anole at La Selva that is likely to bask actively. It is observed most frequently in the laboratory clearing but is known to inhabit the tops of canopy trees elsewhere (Corn 1981).

DIET: This lizard consumes arthropods, especially coleopterans, formicids, dipterans, and araneids (Corn 1981).

REPRODUCTION: Juveniles are more abundant during the wet than the dry season, suggesting increased reproduction during the rainy season (Corn 1981).

REMARKS: It is likely that the anole Perry (1983) described parachuting from limb to limb in a canopy tree was of this species.

This behavior has been described for several individuals observed in canopy and subcanopy vegetation near the Carbono project tower.

NEOTROPICAL CHAMELEON

Polychrus gutturosus

Pls. 108, 109

OTHER COMMON NAMES: Chameleon, Canopy Lizard

SIZE: Largest male 123 mm SVL (n = 2); largest female 170 mm SVL (n = 1) (Roberts 1997).

DISTRIBUTION: Caribbean slopes from Honduras to Colombia; Pacific slopes from central Costa Rica to Ecuador.

IDENTIFYING FEATURES: This is an odd lizard. Phylogenetically, it is related to the anoles because it possesses a homologous dewlap (small in the Neotropical Chameleon) that is used in behavioral displays. The body color can change from leaf green to dark brown. The sides of the face are lime green from the tip of the snout to the eye. Posterior to the eye is a squareish, white patch from the eye to the angle of the jaw. A dark brown, postorbital stripe starts at the eye, extends above the tympanum, and then curves ventrally to the level of the shoulder. Additionally, this lizard typically has a series of diagonal, dark-edged, white to beige bars along the flanks. A heavily dissected, ventrolateral, white stripe may be present from the base of the foreleg to the base of the hind leg. Both sexes possess a dewlap that is dark gray with a series of enlarged, yellow green scales; the dewlap can change to yellow during social encounters (Roberts 1997). The venter and limbs are light lime green, and the tail is light brown with gray blotching. The body is robust but compressed laterally, and the legs are relatively short and thin. A distinctive feature of this lizard is an unusually long tail that often wraps around branches as though prehensile. As the common name suggests, this lizard can change color (Roberts 1997). Additionally, like the true chameleons of Africa and Madagascar, the eyes can be rotated in their sockets independently of each other. Because of these features, this species is not likely to be confused with any other lizard species at La Selva.

SEXUAL DIMORPHISM: The female is larger than the male in other members of this genus (Vanzolini 1983), and data from La Selva appear to corroborate this for the Neotropical Chameleon (Roberts 1997). The male possesses enlarged postanal scales on the ventral surface of the tail immediately posterior to the cloacal opening. Color patterns, especially of the dewlap, need to be examined for sexual dichromatism.

HABITAT: The Neotropical Chameleon lives in trees and probably spends most of its time in the canopy at La Selva. It is diurnal in activity pattern and moves extremely slowly. Relatively little else is know about this lizard because of its arboreal habit, cryptic coloration, and sedentary nature. Most specimens have been collected in trees along or near rivers or streams.

DIET: Members of this genus in South America are known to consume grasshoppers and other large insects and vegetation (Beebe 1944; Duellman 1978; Greene 1986; Vanzolini 1983; Vitt and de la Torre 1996; Vitt and Lacher 1981).

REPRODUCTION: Copulation has been observed at La Selva during May, and palpable eggs were noted in a single female during July, suggesting that reproduction takes place during the wet season (Roberts 1997). Mating is known to occur in trees, where a male climbs on the back of a female and inserts one of his hemipenes into the female's cloaca. Copulation can last several hours, during which the male may switch positions and hemipenes (Roberts 1997). Unlike members of the genus *Norops,* members of the genus *Polychrus* lay many eggs per clutch (seven to 30; Vitt and de la Torre 1996; Vitt and Lacher 1981) in a nest dug in the ground. Hatchling size is unknown, but a single specimen with an umbilical scar from La Selva (CRE 8411) is 57 mm SVL.

REMARKS: This lizard responds aggressively to humans. When approached or harassed it may expand and flatten the body, open the mouth, and bite (Roberts 1997).

Gekkonidae

This family contains the geckos, a group of largely nocturnal lizards that includes over 900 species. The family is widespread, being found on all major continents except Antarctica. These lizards lack eyelids and often possess wide pads under the fingers and toes, which are used to adhere to smooth surfaces. All of these lizards have the curious habit of using their tongues to wipe the clear scale that covers the eye, presumably to clean it. Nearly all members of the family lay one to two eggs per clutch. However, a few genera produce live young. The eggshells can be leathery, or hard like small chicken eggs. All species of the wet forests of Costa Rica have the latter. Geckos are small to midsized lizards that can be distinguished from other lizards by the presence of small and velvety scales on the dorsum and venter. Additionally, geckos have short, wide tails that are no longer than the length of the body and that are round or dorsoventrally flattened. All other lizard families of the lowland wet forests of Costa Rica have tails that are much longer than the rest of the body. The skin of geckos is extremely fragile. Therefore, care should be taken in capturing and handling them to prevent tearing the epidermis. Five species of geckos are present at La Selva, and they represent the fauna known from the rest of the Caribbean lowland wet forests of Costa Rica as well.

Key to the Gekkonidae at La Selva

1a Tail constricted noticeably at base (fig. 16a); fingers and toes webbed . . . . *Thecadactylus rapicauda*

1b Tail not strongly constricted at base (fig. 16b); no webbing between fingers and toes 2

    2a Toes noticeably expanded at tips 3

    2b Toes not noticeably expanded at tips 4

3a Triangular, light spot bordered by dark on dorsum between hind limbs; head and tail not yellow; no black neck collar . . . . *Sphaerodactylus millepunctatus*

3b No triangular, light spot on dorsum between hind legs; head and tail yellow or neck with one or two black collars . . . . *Sphaerodactylus homolepis*

    4a Dorsal color pattern of thin, dorsolateral, light stripes, or a series of light, dorsolateral spots on a dark brown

ground color; light brown dorsal blotches or nuchal collar at back of head . . . .
. . . . *Lepidoblepharis xanthostigma*

4b   Dorsal color pattern either uniform dark gray to black (male) or gray brown with a series of large, black and small, white spots arranged as weak cross bands (juvenile and female); head uniform cinnamon brown (male) or gray with dark brown stripes bordered by small, white spots (juvenile and female) . . . .
. . . . *Gonatodes albogularis*

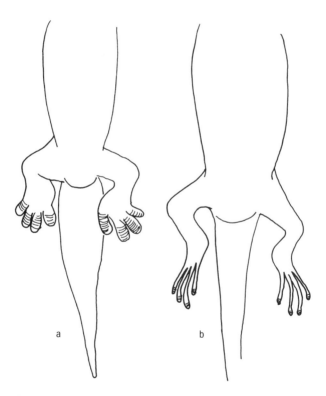

Figure 16. Stylized ventral view of gecko: (a) constricted base of tail; (b) no constriction at base of tail.

YELLOW-HEADED GECKO   *Gonatodes albogularis*
Pls. 110, 111

OTHER COMMON NAMES: Yellowhead Gecko
SIZE: Largest male 44 mm SVL (n = 3);
largest female 37 mm SVL (n = 1).
DISTRIBUTION: Caribbean slopes from Belize to
Venezuela; Pacific slopes from Chiapas, Mexico, to
northern Colombia.

IDENTIFYING FEATURES: The dorsal ground color of the male is dark smoke gray. The tail is similar to the dorsum in color with the exception of its tip, which is white to light gray. The head is cinnamon brown, orange, or yellow and is separated from the body coloration by a bluish white nuchal (neck) bar. The upper lip is cinnamon brown anteriorly with a series of light blue and dark gray spots posteriorly. The chin is uniform cinnamon brown, and the venter of the body and tail is dark smoke gray except for a diffuse, light gray, midventral stripe. The juvenile and female have a very different color pattern. On these animals the ground color is a grayish brown with a series of diffuse, black spots along the middle of the back. A series of small, light punctations line up along the dark spots on the back. Typically, there is a pair of black, dorsolateral spots on the neck, one on each side and bordered by light. These spots separate the coloration of the body from that of the head. The tail is dark brown with a series of indistinct, black rings (diffuse borders). The upper surfaces of the arms and legs are barred with light punctations on a dark gray background. The venter is ashy gray in the throat region and light gray on the belly. In all individuals the iris is cinnamon brown. The adult male cannot be confused with any other species of lizard at La Selva. The juvenile and the adult female are similar in size and general color pattern to the Litter Gecko *(Lepidoblepharis xanthostigma)* and the Spotted Dwarf Gecko *(Sphaerodactylus millepunctatus)*. However, the Litter Gecko has weak stripes instead of cross bands, and the Spotted Dwarf Gecko has a banded tail and distinctly expanded tips to the digits.

SEXUAL DIMORPHISM: The color patterns of the two sexes are distinctive (see above).

HABITAT: This species was first observed at La Selva by Jack Longino, on a wall in the patio area in front of the dining hall (March 16, 1998). The species has since invaded buildings in the laboratory clearing. Elsewhere, this species inhabits areas disturbed by humans. This lizard is often found in and around houses and along fencerows.

DIET: No data on diet are available from La Selva, but this species likely eats small arthropods.

REPRODUCTION: The Yellow-headed Gecko produces a single egg per clutch, but eggs can be laid more or less continuously throughout the year. Incubation requires approximately four months (Leenders 2001).

LITTER GECKO   *Lepidoblepharis xanthostigma*
Pl. 112

SIZE: Largest male 36 mm SVL (n = 10); largest female 34 mm SVL (n = 10).

DISTRIBUTION: Caribbean slopes from Nicaragua to Colombia; Pacific slopes from the Osa Peninsula, Costa Rica, to northern Colombia.

IDENTIFYING FEATURES: The dorsal pattern consists of a dark brown ground color with a pair of dorsolateral, light stripes or series of light spots from behind the eye to the tail. On the neck there is a light bar bordered by dark brown, and the lower lip is barred with light and dark. The venter is striped and/or spotted with light and dark gray anteriorly, becoming salt-and-pepper posteriorly. The tail is dark brown with faint, light spots where the dorsolateral stripes have become dissected. The adult female and juvenile of this species are similar in size and color to those of the Yellow-tailed Dwarf Gecko *(Sphaerodactylus homolepis)*, the Spotted Dwarf Gecko *(S. millepunctatus)*, and the Yellow-headed Gecko *(Gonatodes albogularis)*. However, the Litter Gecko does not have the banded color pattern of the juvenile and female Yellow-tailed Dwarf Gecko and Yellow-headed Gecko and lacks the white spot on the sacrum found in the Spotted Dwarf Gecko.

SEXUAL DIMORPHISM: We know of no external features that distinguish the two sexes.

HABITAT: This species is La Selva's most abundant gecko and one of the most abundant reptiles at this site. However, because it is small and spends most of its time under leaf litter, it is easily overlooked. This lizard can be found wherever leaf litter accumulates and is especially abundant in abandoned cacao *(Theobroma cacao)* plantations.

DIET: The diet consists of small arthropods, dominated by araneids, isopods, and mites, but also including apterygotes, adult and larval dipterans, orthopterans, and hemipterans (Lieberman 1986).

REPRODUCTION: Little is known of reproductive patterns in this species at La Selva. The female produces a single egg in each clutch, and gravid females are observed most frequently during March and April, suggesting a dry season peak in reproduction.

REMARKS: The skin of this lizard tears very easily. Hand capture usually damages the skin.

YELLOW-TAILED DWARF GECKO
Pls. 113–115

Sphaerodactylus homolepis

SIZE: Largest individual 30 mm SVL (n = 4).

DISTRIBUTION: Caribbean slopes from Nicaragua to Panama.

IDENTIFYING FEATURES: This gecko displays striking ontogenetic and intersexual color variation. The juvenile is boldly marked with dark nuchal and axillary bars on a light tan ground color. The head is light gray with a thin, black stripe from the nostril through the eye to the level of the tympanum. The tail is off-white with a series of bold, dark bars. The female has a gray to tan ground color with a series of thin, dark stripes or a series of dark spots that align in rows from the tip of the snout to the level of the armpit (axilla). A dark brown nuchal bar may remain, but other dark bars from the juvenile pattern become brownish gray bordered by indistinct, thin, white bars. The sacrum and hind limbs are heavily mottled with light and dark gray, and the tail is banded with tan on a peach to pink ground color. The venter is uniform light yellow. The male has a series of bold yellow spots on the head arranged in rows from the tip of the snout to the neck. These spots are separated by dark gray. The dorsum of the male is light gray suffused with an indistinct reticulum of darker gray. The tail of the male is bright yellow with a reticulated series of dark gray spots. The color pattern of the adult male is unlike any other lizard species at La Selva. The juvenile and the adult female are similar in size and color pattern to the Litter Gecko *(Lepidoblepharis xanthostigma)*, the Spotted Dwarf Gecko *(S. millipunctatus),* and the Yellow-headed Gecko

(Gonatodes albogularis) (juvenile and female only). However, the Litter Gecko has faint dorsolateral stripes (no bands), the Spotted Dwarf Gecko has a white spot on the sacrum, and the Yellow-headed Gecko has an unbanded tail.

SEXUAL DIMORPHISM: The color patterns described above distinguish the two sexes.

HABITAT: We have observed this species only in buildings.

DIET: Presumably this animal consumes small arthropods.

REPRODUCTION: This species is presumably oviparous with a one-egg clutch.

SPOTTED DWARF GECKO
Pl. 116

Sphaerodactylus millepunctatus

OTHER COMMON NAMES: Spotted Gecko

SIZE: We have no size data from La Selva. However, Leenders (2001) reports a body size of up to 31 mm SVL.

DISTRIBUTION: Caribbean slopes from the Isthmus of Tehuantepec, Mexico, to Costa Rica.

IDENTIFYING FEATURES: This species is easily recognized in the field by the presence of a large, triangular, white spot located above the sacrum; this spot is bordered posteriorly by a V-shaped, dark margin and anteriorly by a dark spot. Additionally, the head is covered by five thin, dark stripes that extend posteriorly to the level of the armpit. The lips are barred with light and dark. This species is similar to the Litter Gecko *(Lepidoblepharis xanthostigma)* and the juvenile and female Yellow-tailed Dwarf Gecko *(S. homolepis)* and Yellow-headed Gecko *(Gonatodes albogularis)*. However, none of these species has a light spot on the sacrum.

SEXUAL DIMORPHISM: We know of no external features that distinguish the two sexes.

HABITAT: This lizard appears to be a relatively rare component of the leaf litter community. We have found it only in abandoned cacao *(Theobroma cacao)* plantations.

DIET: The diet is presumed to consist of small arthropods.

REPRODUCTION: One egg is presumed to compose each clutch.

TURNIP-TAILED GECKO   *Thecadactylus rapicauda*
Pls. 117, 118

OTHER COMMON NAMES: Radish-tail Gecko,
Turnip Tail, Turnip-tail Gecko
SIZE: Largest male 212 mm SVL (n = 2);
largest female 211 SVL (n = 2).
DISTRIBUTION: Caribbean slopes from the Yucatán
Peninsula, Mexico, to Amazonian South America,
including some Antillean islands; Pacific slopes from central
Costa Rica to Ecuador.

IDENTIFYING FEATURES: La Selva's largest gecko, this lizard is also
notable for its extremely wide geographic distribution (see
above). The lizard is flattened dorsoventrally and is easily distinguished from other La Selva geckos by the presence of large, flattened toe disks, webbed hands and feet, and a wide tail that is
constricted at its base. The tail is lost easily, and many individuals
have only a replacement stub; the swollen base of the tail is often
exaggerated in individuals that have regenerated a tail (Vitt and
Zani 1997). The dorsum of this lizard is reticulated with deep
chocolate brown and tan, with numerous black and white or yellow blotches. A postorbital, light stripe occurs on the neck and
continues as a series of white spots to the level of the groin. The
venter is yellow to dirty tan. In the juvenile, the tail is weakly
ringed with an indistinct pattern of light and dark.

SEXUAL DIMORPHISM: The adult male has a more swollen base to
the tail than the female. Limited data on body size have been collected at La Selva; elsewhere the female may be larger than the
male (Vitt and de la Torre 1996).

HABITAT: Although we have found this species under sloughing
bark on dead trees, in large tree holes, and on deep buttresses in
primary forest, it is most easily found in or on buildings, especially the bathroom area at the River Station. This lizard is largely
nocturnal (Vitt and Zani 1997) but may bask on trees near trunk
crevices (Hoogmoed 1973).

DIET: This lizard consumes large, nocturnal arthropods, especially roaches, crickets, beetles, and moths (Hoogmoed 1973; Vitt
and de la Torre 1996; Vitt and Zani 1997). It may be beneficial for
roach control in station buildings.

REPRODUCTION: Patterns of reproduction are unknown at La
Selva, but elsewhere, this species lays a single egg per clutch, with

clutches produced during several months of the year (Duellman 1978; Hoogmoed 1973; Vitt and de la Torre 1996; Vitt and Zani 1997). The eggshell is calcified, creating a hard outer coat.

REMARKS: When descending from an arboreal site this animal may jump, spread its toes, and glide downward (Vitt and de la Torre 1996). This lizard is social and communicates by vocalization. The call is a series of 15 to 20 "chack" notes given in rapid succession (Hoogmoed 1973) and may be used to establish a territory (Kluge 1967). When disturbed this lizard raises its tail and waves the tip; this behavior is presumed to be used to distract predators, but the tail also may be lost during aggressive interactions (Vitt and Zani 1997).

Xantusiidae

The common name for members of this family is night lizards, because some species are thought to be crepuscular or nocturnal. This is a small family of 17 species distributed in the desert southwest of North America, Central Mexico, Central America, and Cuba. The family is characterized by enlarged, beaded scales on the dorsum and large, rectangular scales on the venter. Members of the family also have no moveable eyelid. Females retain eggs in the uterine tract and give birth to live young. In the lowland wet forests of Costa Rica, these lizards are secretive animals that live under logs and debris. La Selva has a single species from this family, and it represents the fauna known from the rest of the Caribbean lowland wet forests of Costa Rica as well.

TROPICAL NIGHT LIZARD　　*Lepidophyma flavimaculatum*
Pls. 119, 120
OTHER COMMON NAMES: Yellow-spotted
Night Lizard
SIZE: Largest individual 97 mm SVL (n = 10).
DISTRIBUTION: Caribbean slopes from Veracruz,
Mexico, to Panama.
IDENTIFYING FEATURES: This odd lizard is characterized by large, beadlike dorsal scales (surrounded by and rising above small, rounded scales) and square ventral ones. The body is black to rusty brown with a series of large, yellow spots, which are often faded and have indistinct borders. The lower jaw is robust, giving

the head a distinctive blocky appearance, and has a series of vertical light (white or yellow) and dark (black) bars. The upper lip is
black with a series of light spots. The top of the head is uniform
black, the venter is uniform dark brownish gray, and the tail is
dark gray to black with small, indistinct, light spots. Enlarged
caudal scales form whorls around the tail; each whorl is separated
by about three rows of smaller scales. The iris is dark gray. No
other lizard at La Selva has beaded scales and a dark body with
light spots.

SEXUAL DIMORPHISM: We know of no external features that distinguish the two sexes.

HABITAT: This species is difficult to observe because it resides in
large, rotting logs, stumps, crevices in tree trunks, or leaf litter in
tree buttresses. It can sometimes be observed during night walks
by careful surveys of these microhabitats or by turning over old
trail boards piled in primary forest.

DIET: At La Selva, this species is known to consume formicids and
centipedes (Lieberman 1986). Slevin (1942) reports this lizard
consuming isopterans in Guatemala.

REPRODUCTION: The Tropical Night Lizard is unusual in its reproductive habits because it is gonochoristic in the northern portion
of its range and parthenogenetic in the southern portion. La
Selva's population is presumed to be parthenogenetic, but approximately 25 percent of the population is male (Bezy 1989).
This species is also viviparous with true placentation: it has a
chorioallantoic placenta, as mammals do. Many juveniles were
observed July 20–25, 2001, suggesting an early wet season birth-
ing period.

Scincidae

With approximately 1,100 species, this is the second largest family of squamates, surpassed only by the colubrid snakes. These
lizards are found on all major continents, except Antarctica, and
are found in virtually every habitat. Additionally, they come in a
variety of shapes, ranging from forms with wide bodies, thick
heads, and short, fat tails, to elongate forms with limbless, snakelike bodies. Most species are diurnal insectivores that can be distinguished from other squamates by the presence of smooth,
cyloid (half-moon-shaped) scales on the dorsum and venter,
combined with a tongue that is split at the tip but not deeply

forked like a snake's tongue. These lizards taste their environment with frequent flicks of their tongues, much as snakes do. Two skink species are present at La Selva, and they represent the fauna known from the rest of the Caribbean lowland wet forests of Costa Rica as well.

Key to the Scincidae at La Selva

1a Body with lateral, white stripe (half a scale row wide) confluent with supralabials and extending from axilla to groin
. . . . *Mabuya unimarginata*

1b Body without wide, lateral, light stripe from axilla to groin
. . . . *Sphenomorphus cherrei*

BRONZE-BACKED CLIMBING SKINK

Pl. 121

Mabuya unimarginata

OTHER COMMON NAMES: Lucia, Central American Mabuya

SIZE: Largest individual 77 mm SVL (n = 1).

DISTRIBUTION: Caribbean slopes from Veracruz, Mexico, to Panama; Pacific slopes from Sinaloa, Mexico, to Ecuador.

IDENTIFYING FEATURES: The middorsum of this lizard is a coppery brown with black spots; this color pattern covers the top of the head as well. Starting from the tip of the snout, a wide, black stripe extends along the side of the face below the eye, passes above the pectoral and pelvic limbs, and continues along the side of the tail. This stripe is one to two scale rows wide. Below this black stripe is a white stripe that starts at the supralabials and extends through the axilla to the level of the groin. Below the white stripe is a dark brown stripe, two to three scale rows wide, from the angle of the jaw to the groin; this dark stripe separates the brown dorsal coloration from the light cream ventral coloration. The bottoms of the hands and feet are dark brown, contrasting with the cream coloring of the arms and legs. This species can be confused with the Litter Skink *(Sphenomorphus cherrei)* and the Rainforest Celestus *(Celestus hylaius)*. However, both of these lizards lack the distinct lateral, light stripe of the Bronze-backed Climbing Skink.

SEXUAL DIMORPHISM: We know of no external features that distinguish the two sexes.

HABITAT: Of the two species of skinks at La Selva, this one is less frequently seen, presumably because it is arboreal. We have seen only one specimen and, therefore, do not know the habitat affinities of this species.

DIET: This species presumably is insectivorous.

REPRODUCTION: Reproduction is unusual in that eggs are retained in the oviduct, and the female gives birth to live young. From four to nine offspring are produced in each litter (Alvarez del Toro 1983; McCoy 1966; Webb 1958).

LITTER SKINK   *Sphenomorphus cherrei*
Pl. 122

OTHER COMMON NAMES: Lucia, Escincela Parda, Brown Forest Skink

SIZE: Largest male 53 mm SVL (n = 3); largest female 56 mm SVL (n = 8).

DISTRIBUTION: Caribbean slopes from Veracruz, Mexico, to Panama; Pacific slopes from Costa Rica to Panama.

IDENTIFYING FEATURES: The dorsum of this species is a dark gray brown. A wide, dark stripe begins at the posterior margin of the eye and continues to near the level of the groin, becoming indistinct posteriorly. From the posterior margin of the ear to the level of the armpit, the dark stripe is bordered below by a series of light yellow spots on a light gray ground color. The venter is uniform grayish yellow, and the iris is dark brown to black. This species looks similar to the Bronze-backed Climbing Skink (*Mabuya unimarginata*) but differs from that species by lacking a bold, white, lateral stripe. The Litter Skink could also be confused with the Rainforest Celestus (*Celestus hylaius*), but the latter species is green.

SEXUAL DIMORPHISM: We know of no external features that distinguish the two sexes.

HABITAT: This lizard is an abundant member of the leaf litter community. It can be found in primary and secondary forest, where it slithers along the surface momentarily and then disappears under the litter. The species is particularly abundant in abandoned cacao (*Theobroma cacao*) plantations.

DIET: The diet consists of a variety of arthropods (orthopterans, hymenopterans, dipterans, homopterans, coleopteran larvae,

isopods, centipedes), especially areneids and hemipterans (Lieberman 1986).

REPRODUCTION: The female typically produces two eggs per
clutch, and these are deposited in the leaf litter. Reproduction can
take place throughout the year (Fitch 1970; Greene 1969) but increases during the wet season (Fitch 1973a).

REMARKS: The generic designation has vacillated between *Sphenomorphus* and *Leiolopisma*, but currently *Sphenomorphus* is fa-
vored.

Teiidae

The family Teiidae contains 105 species of small to midsized
lizards that are sun-seeking, active foragers. The family is found
throughout most of the United States, Mexico, and Central
America, and its range extends through all but the southernmost
region of South America. The family is characterized morpho-
logically by the presence of small, velvety scales on the dorsum
and enlarged, rectangular scales on the venter. All members of the
family lay eggs in ground nests. These nests are communal in
some species. In the lowland wet forests of Costa Rica, these
lizards are found along beaches, in open pastures, along trails,
and in treefall gaps. Only one species is present at La Selva. One
additional species is known from elsewhere in the Caribbean
lowland wet forests of Costa Rica.

CENTRAL AMERICAN WHIPTAIL   *Ameiva festiva*
Pls. 123–125

OTHER COMMON NAMES: Ameiva, Chisbala,
Middle American Ameiva, Tropical Whip-
tailed Lizard

SIZE: Largest male 118 mm SVL (n = 41);
largest female 110 mm SVL (n = 19).

DISTRIBUTION: Caribbean slopes from Tabasco, Mexico, to
northern Colombia; Pacific slopes from San José, Costa Rica,
to northwestern Colombia.

IDENTIFYING FEATURES: The juvenile has a dark brown dorsal
ground color with a bold, middorsal, light yellow to white stripe.
A pair of thin, heavily dissected, dorsolateral, light stripes is found

on each side of the body, and the tail is metallic light blue. The middorsal stripe fades in large adults, leaving a wide, light brown stripe that is bordered by the heavily dissected dorsolateral, light stripes on the sides of the body. These stripes are yellow anteriorly shading to orangish posteriorly. The blue tail coloration of the juvenile fades in the adult. No other lizard at La Selva is as fast or has velvety scales above and rectangular ones below.

SEXUAL DIMORPHISM: The male has a burnt orange throat region, whereas that of the female is turquoise. We found no noticeable dimorphism in snout-to-vent length for this species at La Selva; however, the adult male weighs more and develops a larger head than the female (Vitt and Zani 1996).

HABITAT: This is the common "trail lizard" of La Selva. As with all members of the family Teiidae, the Central American Whiptail is active on sunny days, when it can be seen basking in sun flecks and then dashing through darker areas until the next sun fleck is sought. This shuttling between sun and shade maintains the body temperature at a high, constant level (van Berkum 1986). This lizard creates burrows under trail boards and other woody debris; burrows serve as retreats on cloudy days and at night.

DIET: The diet is catholic, including arthropods (often orthopterans, blattids, isopods, and araneids; Hillman 1969; Lieberman 1986; Vitt and Zani 1996), as well as small vertebrates. We have observed this species eating an adult Bransford's Litterfrog *(Eleutherodactylus bransfordii)* and a Ground Anole *(Norops humilis)* at La Selva.

REPRODUCTION: We have reproductive data from a single female from the Braulio Carrillo National Park extension that had four oviductal eggs. However, published records suggest that breeding occurs year-round and that the female lays from one to five (usually two) eggs (Fitch 1970; Smith 1968; Vitt and Zani 1996) in a nest dug into the ground or under leaf litter.

REMARKS: A second species of *Ameiva,* the Four-lined Whiptail *(A. quadrilineata),* is known from coastal areas and sites disturbed by human activities along the Caribbean slopes of Costa Rica. No voucher of this species is known from La Selva. However, it may be present or may invade the site because of the encroachment of human activities. Any *Ameiva* that does not conform to the species description for the Central American Whiptail should be compared with the description for the Four-lined Whiptail (see under "Additional Species").

Anguidae

Members of this family are elongate, sometimes limbless, shiny lizards that often undulate when they move, like snakes. The family contains approximately 110 species that are found on all continents except Africa, Australia, and Antarctica. One center of diversity of this family is the wet and cloud forests of Central America. Here, these lizards are terrestrial or arboreal. Some forms lay eggs, whereas others retain eggs in the uterine tract and give birth to live young. The members of the family inhabiting the lowland wet forests of Costa Rica are characterized by large, rectangular, keeled scales. The region between the keels on each scale is flat, which makes the lizards shine like skinks. However, the tongue of anguids is deeply forked, unlike the more rounded tongue of skinks. Anguids use this tongue to taste their environment, in the same manner that snakes do. Three species of anguids occur at La Selva, all of which are secretive and, therefore, poorly known. These three species represent the fauna known from the rest of the Caribbean lowland wet forests of Costa Rica as well.

Key to the Anguidae at La Selva

1a Sides of body with bold, vertical, light and dark bars 2
1b Sides of body more or less uniform . . . . *Celestus hylaius*
  2a Middorsum with wide, dark gray bars bordered posteriorly by thin, light bars; sides of body gray with tomato red bars . . . . *Diploglossus monotropis*
  2b Middorsum uniform; sides of body with light and dark bars (lost in large adults) . . *Diploglossus bilobatus*

RAINFOREST CELESTUS   *Celestus hylaius*
Pls. 126, 127
SIZE: Largest individual 106.8 mm SVL
(n = 3; Savage and Lips 1993).
DISTRIBUTION: Caribbean slopes of Costa Rica.
IDENTIFYING FEATURES: This species looks like a skink in that the scales are smooth and shiny. The top of the head is greenish brown shading to coppery brown on the dorsum. Many scales are tipped with black, giving the dorsum a salt-and-pepper appearance. These black markings coalesce along the sides of the body, forming an indistinct, lateral, dark

stripe from behind the eye to the groin. The labial scales are bright yellowish green, as is the venter. This species is vaguely similar to the Litter Skink *(Sphenomorphus cherrei)* and the Bronze-backed Climbing Skink *(Mabuya unimarginata)*. However, these species are uniform brownish in color, and the Bronze-backed Climbing Skink has a distinct, light, lateral stripe.

SEXUAL DIMORPHISM: We know of no external features that distinguish the two sexes.

HABITAT: This species is rarely observed, perhaps because it is largely arboreal. It is occasionally found in houses (O. Vargas 1986, personal communication).

DIET: This lizard is presumed to eat arthropods.

REPRODUCTION: This species is presumed to be viviparous, with a relatively small litter size.

REMARKS: The type specimen for this recently described species is an individual from La Selva; previous literature may refer to it as *C. cyanochloris* (Savage and Lips 1993).

TALAMANCAN GALLIWASP   *Diploglossus bilobatus*
Pl. 128

OTHER COMMON NAMES: Galliwasp

SIZE: The only specimens from La Selva that we have seen are three juveniles captured during the leaf litter project described by Lieberman (1986). Elsewhere, this lizard reaches 92 mm SVL (Myers 1973).

DISTRIBUTION: Caribbean slopes of Costa Rica and Panama; Pacific slopes of central Costa Rica.

IDENTIFYING FEATURES: The dorsum of this species is dark brown, turning light brown rather abruptly on the sides. The dark brown and light brown are separated by a thin, indistinct, dorsolateral, dark stripe. A key field characteristic for this species is a series of white to light green spots from the tip of the snout, along the supralabials, to the anterolateral aspect of the trunk. These spots may be associated with a series of lateral, dark bars that extends the length of the trunk. The spots may become faded or lost in large adults. The head is greenish brown, and the venter is gray to pinkish. The most similar other species at La Selva is the Coral-mimic Galliwasp *(D. monotropis)*. However, the latter species has red markings along the sides that are not found in the Talamancan Galliwasp.

SEXUAL DIMORPHISM: We know of no external features that distinguish the two sexes.

HABITAT: We know of no description of habitat selection in this lizard.

DIET: Nothing is known of the diet of this species, but it is presumed to eat arthropods.

REPRODUCTION: This species lays eggs (six) in a ground nest that is attended by the female (Fitch 1970).

CORAL-MIMIC GALLIWASP   *Diploglossus monotropis*
Pl. 129

OTHER COMMON NAMES: Escorpion Coral, Madre de Culebra, Madre Coral, Galliwasp

SIZE: A single specimen from La Selva measured ca. 165 mm SVL (R. Coleman 1992, personal communication). Elsewhere, this species is known to reach 215 mm SVL (Myers 1973).

DISTRIBUTION: Caribbean slopes of Nicaragua to northern Colombia; Pacific slopes from Panama to Ecuador.

IDENTIFYING FEATURES: The dorsal pattern consists of a dark gray to black ground color changing abruptly to tomato red along the sides. The body has a series of thin bands that are bright yellow (one to two scale rows) edged anteriorly by black (one to two scale rows) that extend across the middorsum and sides. The light bands often are offset at the middorsal line. The head is gray with diffuse, dark blotches, and the iris is dark, rusty red. The supralabials and infralabials are bright yellow with a series of thin, black bars, creating a banded pattern. The venter is yellow to tomato red, with the coloration wrapping about halfway up the sides of the body, but much less than halfway up on the tail. The tail is dark gray with a series of yellow bands edged anteriorly by black. These yellow bands are wider (two to three scale rows) than those along the body. The overall pattern of the body mimics tricolored coralsnakes, whereas the pattern of the tail mimics bicolored coralsnakes. At La Selva, this species is most similar to the Talamancan Galliwasp *(D. bilobatus)*. However, the Talamancan Galliwasp lacks red, lateral coloration.

SEXUAL DIMORPHISM: We know of no external features that distinguish the two sexes.

HABITAT: This is a large, terrestrial anguid and a rare find. The only specimens that we know of from La Selva were found in second-growth forest at about 1,500 m along the SSE trail (O. Vargas 1992, personal communication) and on the steps leading from the River Station to the boat landing on the Rio Puerto Viejo (R. Coleman 1992, personal communication).

DIET: Nothing is known of the diet, but this lizard presumably consumes invertebrates. Leenders (2001) suggests that this species feeds on land crabs.

REPRODUCTION: This species is presumed to lay eggs.

REMARKS: When disturbed, this lizard slithers along the ground like a snake.

Boidae

This family contains the large, heavy-bodied constrictors. Approximately 65 species compose this family, and they are distributed on all major continents except Antarctica. Some of these snakes are impressively large, and the largest of all squamates, the reticulated python (approximately 10 m), is in this family. However, most members are less than 2 m in length. Members of this family can be distinguished from most other snake families by the presence of small scales covering the top of the head; smooth, shiny scales across the dorsum; and narrow belly scutes. In the lowland wet forests of Central America, these snakes are slow-moving animals that may be found on the ground or high in vegetation. Two species of boids occur at La Selva, and they represent the fauna known from the rest of the Caribbean lowland wet forests of Costa Rica as well.

Key to the Boidae at La Selva

1a Labial pits present on posterior-most upper and lower labial scales (fig. 17); body with muted brown blotches on a reddish to orangish brown ground color . . . .
. . . . *Corallus annulatus*

1b No labial pits; body with bold, dark blotches. . . . .
. . . . *Boa constrictor*

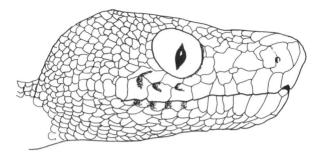

Figure 17. Lateral view of head of *Corallus annulatus* showing position of labial pits (stippled areas on posterior labial scales).

BOA CONSTRICTOR
Pl. 130

Boa constrictor

OTHER COMMON NAMES: Bequer, Boa
SIZE: Largest male 2,220 mm TL (n = 2); largest female 3,040 mm TL (n = 11).
DISTRIBUTION: Caribbean slopes from Tamaulipas, Mexico, through Amazonian South America to Argentina; Pacific slopes from Sonora, Mexico, to northern Colombia.

IDENTIFYING FEATURES: This large, heavy-bodied snake is unmistakable. The dorsum is grayish brown with dark brown blotches, each bordered with black. Each blotch also has paravertebral, yellow spots. The lateral portion of the body has reddish brown ocelli, each with a yellow center. The tail has a series of reddish ocelli bordered by black and off-white. The venter is gray with black spots. Each scale is flat and smooth, giving the animal an iridescent sheen. The genus *Corallus* has a body shape similar to that of the genus *Boa,* but the former genus is smaller, has more muted colors, and possesses a series of heat-sensitive pits on the lower labial scales.

SEXUAL DIMORPHISM: Both the male and the female possess a pair of spikelike structures at the posterolateral portion of the body near the vent, but these are better developed (longer and thicker) in the male. These remnants of the pelvic girdle are easy to overlook. The adult female is larger than the adult male (S. M. Boback 2004, personal communication).

HABITAT: The Boa is largely nocturnal and can be found in primary and secondary forest at La Selva, including cleared areas. It occurs both on the ground and in trees. Several have been observed at night in the buildings, presumably hunting rats.

DIET: The Boa Constrictor eats a variety of vertebrate prey including lizards (Green Iguana *[Iguana iguana]* and Spiny-tailed Iguana *[Ctenosaura similis]*), birds (poultry, Blue-gray Tanager *[Thraupis episcopus]*, antbird), and mammals (opossums *[Didelphis* spp.*]*, bats, spiny rats *[Proechimys* spp.*]*, Agouti *[Agouti paca]*, Tree Porcupine *[Coendou rothschildi]*, rabbits, deer, White-nosed Coati *[Nasua narica]*, Ocelot *[Felis pardalis]*, mongoose; Greene 1983). To this list we add an Orange-billed Sparrow *(Arremon aurantiirostris)* consumed by a juvenile and a rabbit eaten by an adult. Prey are killed in powerful constricting coils.

REPRODUCTION: The female bears live young (viviparous). One caged adult at La Selva gave birth to 52 offspring (at 1 A.M. on July 23, 1983); elsewhere this species is known to mate from August to March and give birth to 20 to 67 young from March through August (Greene 1983).

ANNULATED TREE BOA

Corallus annulatus

Pls. 131, 132

SIZE: Largest individual 1,685 mm TL (n = 3).

DISTRIBUTION: Caribbean slopes from Guatemala to Colombia; Pacific slopes of Colombia and Ecuador.

IDENTIFYING FEATURES: This handsome boid has two color patterns. Young individuals have a dorsal ground color that is light reddish to orangish brown with a series of large blotches along each side; each blotch is lighter reddish brown bordered by dark brown. Older individuals are gray with ocelli along each side. Each ocellus is light brown surrounded by a wide, dark brown border. Most ocelli are offset from one side to the other, creating a zigzag pattern of dark brown along the middorsum. However, a few ocelli may meet at the midline, creating dark, saddle-shaped markings across the back. The head is triangular and is dark brown with a reticulum of black spots, lines, and U-shaped markings. The labials have heat-sensitive pits that are visible on the upper and lower lips (fig. 17). The venter is light reddish or purplish brown with small, light and dark spots. The tongue is dark brown with a light gray tip, and the iris is grayish brown. This snake is not likely to be confused with any other.

SEXUAL DIMORPHISM: Both the male and female possess a pair of spikelike structures at the posterolateral portion of the body near the vent; these are better developed (longer and thicker) in the male.

HABITAT: This snake is arboreal, but because few specimens have been seen at La Selva, little is known about its habitat preferences. We suspect that it occupies the canopy trees of primary forest.

DIET: Few diet observations have been made for this species anywhere. However, we isolated hummingbird feathers from the feces of a juvenile at La Selva.

REPRODUCTION: Like all New World boas, the Annulated Tree Boa gives birth to live young (viviparous).

REMARKS: This snake is generally calm when captured but occasionally may gape, as though threatening to bite. A species similar to the Annulated Tree Boa, the Rainbow Boa *(Epicrates cenchria)*, may be present at La Selva and is found at other lowland sites along the Caribbean slopes of Costa Rica. The Rainbow Boa keys to the Annulated Tree Boa in our key. Therefore, individuals should be compared against the description of the Rainbow Boa (see under "Additional Species").

Ungaliophiidae

This family contains the three species of Neotropical dwarf boas. Although long grouped with the family Tropidophiidae, we follow Zaher (1994) in placing these snakes in a separate family because their historical affinities are with the family Boidae and not with Colubridae and Viperidae (as in Tropidophiidae). Members of the family Ungaliophiidae are found in wet forests along the Caribbean slopes from Chiapas, Mexico, through Panama to northern Colombia, and along the Pacific slopes from Oaxaca, Mexico, to southwestern Costa Rica. They are like boids in that remnants of the pelvic girdle are present. However, this group appears like colubrids in that the body is relatively thin, the head is covered with enlarged, symmetrical, platelike scales (not small scales as in boids), and the venter consists of wide scutes (not narrow ones). These snakes are highly arboreal and are thought to be associated with thick epiphytic growths on canopy trees.

Only one species is present at La Selva, and this is the only species found in other Caribbean wet forest areas of Costa Rica.

CENTRAL AMERICAN DWARF BOA
Pl. 133

Ungaliophis
panamensis

SIZE: We have no measurements of this species from La Selva; elsewhere this snake reaches sizes as large as ca. 750 mm TL (Obst et al. 1988).

DISTRIBUTION: Caribbean slopes from Nicaragua to northern Colombia; Pacific slopes from central Costa Rica to Panama border.

IDENTIFYING FEATURES: The color pattern of this snake is unforgettable. The dorsum has a light brown to gray ground color with a series of paired, dark brown to black spots along the middorsum. Each spot is round to triangular in shape and is encircled by a thin, white or yellowish border, the outer edge of which may be bordered by brick red. A second series of indistinct, dark spots is found along the lateral surface. The head is brown to gray with an elongate, upside-down, V-shaped, dark brown to blackish blotch, with the apex of the V pointed toward the snout. The rounded tip of the V begins on the snout and the "arms" extend to the neck, where their termini are rounded. The entire head blotch is edged in white or yellow. A thin, dark brown stripe runs from each eye to at least the angle of the jaw. The venter of the body is gray brown with dark spots, whereas the chin and the venter of the tail are gray. No other snake at La Selva has bold, large, dark spots along with a thick, muscular body.

SEXUAL DIMORPHISM: Both the male and the female possess a spurlike projection on each side of the body near the vent; these are larger (longer and thicker) in the male.

HABITAT: This snake is arboreal and probably burrows in the epiphytes of large canopy trees (Corn 1974). At La Selva this species has invariably been found associated with large tree falls that contained heavy epiphyte loads. It has been observed in both primary and secondary forest.

DIET: We know of no description of diet for this species.

REPRODUCTION: In captivity the male bites the tail of the female during copulation, rather than the head or neck like most squamates (Burge 1995). Little else is known about reproduction in this species.

Viperidae

This is the family of the pitvipers and true vipers. This family contains approximately 220 species that are distributed on all major continents except Australia and Antarctica. New World members are characterized by the presence of a heat-sensitive pit on the side of the face between the eye and the external naris (fig. 13a). The family also is characterized by elongate, mobile fangs for delivering venom; vertically elliptical pupils; and strongly keeled scales (which give the skin a rough appearance). However, if you are close enough to the animal to use these characteristics for identification purposes, you are probably too close. The venom glands are located at the back of the head, contributing to the head's triangular shape. This is a good field characteristic, but several colubrid snakes flare the bones at the back of the skull to mimic this head shape. Four species of viperids inhabit La Selva. For many years these snakes were placed within two genera, *Lachesis* (for the Bushmaster) and *Bothrops* (for all others). However, the generic designations used here reflect recent systematic arrangements by Campbell and Lamar (1989), Crother et al. (1992), and Werman (1992) in which additional genera are recognized in order to make *Bothrops* monophyletic.

Key to the Viperidae at La Selva

1a Rostral scale enlarged and upturned; body noticeably short and wide (stocky). . . . . *Porthidium nasutum*
1b Rostral scale neither enlarged nor upturned; body not noticeably short and wide (stocky). 2
  2a Supraocular scale with elongate spines (fig. 13a); dorsum mottled lichen green or uniform brilliant yellow . . . . *Bothriechis schlegelii*
  2b Supraocular scales without elongate spines; coloration other than green or brilliant yellow 3
3a Lateral, triangular markings darker (brown to black) than grayish brown, middorsal diamonds; lateral triangles separated from dorsal diamonds by white edging . . . . *Bothrops asper*
3b Lateral, triangular markings lighter (reddish tan to brown) than brown to black, middorsal diamonds; lateral triangles not separated from dorsal diamonds by white edging. . . . . *Lachesis stenophrys*

EYELASH VIPER
Bothriechis schlegelii

Pls. 134, 135

OTHER COMMON NAMES: Oropel, Bocaraca, Toboba Pestanas, Eyelash Pitviper, Schlegel's Viper, Palm Viper, Eyelash Palm-pitviper

SIZE: Largest male 494 mm TL (n = 2); largest female 698 mm TL (n = 3). Leenders (2001) reports that this species reaches sizes up to 800 mm TL.

DISTRIBUTION: Caribbean slopes from Chiapas, Mexico, to north-western Venezuela; Pacific slopes from central Costa Rica to Ecuador.

IDENTIFYING FEATURES: This species is of medium size and has two dramatically different color morphs at La Selva. One is moss-patterned, consisting of mottled greens, browns, and yellows; the other is a brilliant, uniform yellow. The green morph is encountered far more frequently than the yellow morph. This species is unique in possessing elongate, eyelashlike scales (superciliaries) that give the snake its English common name, the Eyelash Viper. This feature distinguishes the moss-patterned morph from the Lichen-colored Slugeater *(Sibon longifrenis)*, a colubrid with an apparently mimetic color pattern. No snake at La Selva resembles the yellow morph.

SEXUAL DIMORPHISM: We know of no external features that distinguish the two sexes.

HABITAT: This species is La Selva's only arboreal viper (excepting the Terciopelo *[Bothrops asper]*, which on rare occasions takes to the trees). It can be found coiled on vegetation, often sandwiched between overlapping understory palm leaves (e.g., *Geonoma* spp.) or on lianas and vines. It is nocturnal and can be found in closed-canopy secondary and primary forest.

DIET: The Eyelash Viper is known to consume lizards, frogs, snakes, and small birds and mammals as its major prey items (Alvarez del Toro 1983; Bridegam et al. 1990; Amaral 1927). The tip of the tail of the juvenile is bright yellow and may be wiggled to lure lizards and frogs that are then consumed (Neill 1960). Small prey often are held in the mouth after the predatory strike, an unusual behavior among pitvipers (Boyer et al. 1991).

REPRODUCTION: This species gives birth to live young, but little is known of litter size or the timing of reproduction. A litter containing 18 offspring has been reported from Costa Rica (Picado 1976).

REMARKS: The venom of this snake is potent, and human fatalities from bites occur yearly in Costa Rica (Seifert 1983). One feature of the natural history of this species demands examination. The color morphs are distinctive and apparently genetic in origin. The yellow morph has been observed to perch in wild banana plants (*Heliconia* spp.) in a fashion that suggests that the yellow color could serve to attract hummingbirds that pollinate those plants. At least one photo documents such a predatory attempt (Greene 1997). The mottled green morph is frequently encountered on lichen- and moss-covered leaves. However, nonrandom habitat use by the two morphs remains to be quantitatively documented.

TERCIOPELO
Bothrops asper

Pl. 136

OTHER COMMON NAMES: Toboba Real,
Fer-de-Lance, Barba Amarilla
SIZE: Largest male 1,300 mm TL (n = 1);
largest female 1,595 mm TL (n = 4). Elsewhere
this species reaches sizes as large as 2,500 mm TL
(Campbell and Lamar 1989), and some individuals at La
Selva are likely this large as well.
DISTRIBUTION: Caribbean slopes from Tamaulipas, Mexico, to
Venezuela; Pacific slopes from Costa Rica to Ecuador.
IDENTIFYING FEATURES: The adult Terciopelo is a large snake with a
dorsum that consists of a matrix of dark (brown to black) and
light (tan to gray) markings. The color pattern is extremely cryptic when viewed against a background of leaf litter. The Terciopelo differs from the Bushmaster (*Lachesis stenophrys*) in having tan to gray, diamond-shaped marks that are bordered by light
as the major color pattern of the middorsum. Separating the light
diamonds are dark brown to black triangles along each side, the
apices of which may meet at the middorsal line, creating X-shaped patterns. The juvenile has a color pattern that is similar to
the adult with the exception of a yellow tail tip in the neonate
male. Only one other species is likely to be confused with a Terciopelo; this is the rear-fanged colubrid, the False Terciopelo
(*Xenodon rabdocephalus*), a species with a color pattern that
mimics the Terciopelo. The Terciopelo can be distinguished from

the False Terciopelo by the presence of a vertical pupil, like a cat's eye, in the Terciopelo rather than a round one as in the False Terciopelo.

SEXUAL DIMORPHISM: We know of no external features that distinguish the two sexes other than the yellow tail tip in the juvenile male.

HABITAT: This species is probably the most commonly encountered pitviper and is one of the most abundant snakes at La Selva. It is largely nocturnal and can be found in virtually all habitats. The belly scales of this snake when coiled reflect bright white when spotlighted, therefore the snake can be detected relatively easily by those using flashlights to carefully survey their path. The juvenile frequently, and the adult occasionally, perches in low vegetation at night. During the day it often coils in a secluded, terrestrial refugium where it might not be noticed until the observer is at a shockingly close distance. Fortunately, such an individual is usually asleep and requires a major disturbance to be forced into action.

DIET: The adult typically preys on mammals as large as opossums (*Didelphis* spp.); the juvenile consumes lizards, frogs, snakes, and invertebrates (Buttenhoff and Vogt 1995, 1997; Greene 1992; Greene and Hardy 1989; Solórzano and Cerdas 1989). The yellow tail tip in the juvenile male is thought to serve as a caudal lure for these prey.

REPRODUCTION: The female gives birth to live young. The litter size is exceptionally large, with up to 86 offspring (Hirth 1964), perhaps explaining the abundance of this species. Mating occurs in March, and young are born from September to November (Solórzano and Cerdas 1989).

REMARKS: The adult Terciopelo can move rapidly and unpredictably when molested. It strikes readily and should be treated with respect.

BUSHMASTER

Lachesis stenophrys

Pls. 137, 138

OTHER COMMON NAMES: Matabuey, Cascabel Muda, Central American Bushmaster

SIZE: Largest male 2,065 mm TL (n = 7); largest female 1,800 TL (n = 4).

DISTRIBUTION: Caribbean slopes from Nicaragua to Colombia; Pacific slopes from Panama to Ecuador.

IDENTIFYING FEATURES: This large snake is unrivaled in elegance and is the subject of much jungle lore. The dorsal color pattern consists of shades of brown, tan, and yellow. The middorsum usually has a series of dark brown, diamond-shaped markings, the centers of which can be reddish tan, except for the middorsal-most two to three scale rows. Here, the scales are large, beaded, and dark brown to black; this creates a middorsal, dark stripe that is interrupted by the white patches that separate consecutive diamonds. The sides of the body consist of tan, triangular markings created by the lateral extensions of the dark, middorsal diamonds. A dark stripe extends from the eye to the back of the lateral surface of the head. This separates the dark, dorsal head coloration from the light, ventral portion. On some individuals the dark, diamond-shaped markings are lightened or absent, creating a color pattern of light tan with a middorsal, dark stripe. The color pattern of the Bushmaster is vaguely similar to that of the Terciopelo _(Bothrops asper)_, but the latter species lacks the beaded, middorsal scales and the dark, middorsal stripe. No species at La Selva mimics the color pattern of the Bushmaster.

SEXUAL DIMORPHISM: We know of no external features that distinguish the two sexes.

HABITAT: A resident of primary forest, this snake is most often encountered coiled on the surface of leaf litter in relatively secluded spots. Its cryptic color pattern makes it extremely difficult to detect. The Bushmaster is largely active at night and seeks secluded sleeping sites during the day.

DIET: The most common diet item at La Selva is spiny rats _(Proechimys_ spp.; Greene 1988), which the snake captures by coiling next to mammal trails and waiting in ambush (Greene and Santana 1983). The Bushmaster is unusual among pitvipers in that the food items are small relative to the size of the snake, and are held in the mouth after the strike, rather than being released and trailed (Boyer et al. 1991).

REPRODUCTION: This is the only egg-laying pitviper in the New World. Clutch size is thought to be 10 to 12 eggs. Reports of egg-deposition dates are varied, suggesting no set reproductive season (Fitch 1970).

REMARKS: Two bits of lore about the Bushmaster need substantiation from field data. First is the suggestion that the Bushmaster is observed unusually frequently next to *Welfia* spp. palms. Because spiny rats are the putative dispersal agent of this palm (Vandermeer et al. 1979) and because pitvipers hunt by following prey odor trails to areas of high concentration and then sitting and waiting for the prey, the local lore may be true. If the Bushmaster regulates the dispersal agent of *Welfia* spp., an important understory plant species, then this predator might play a role in determining forest structure. The second bit of lore is that the aggressive reactions of the Bushmaster toward humans are clustered during the early dry season (typically February) when the female defends her nest. To date, scant data are available from anywhere within the range of the Bushmaster to document the reproductive season and associated behaviors. We follow Zamudio and Greene (1997) in separating this species from the Bushmaster found in South America, *L. muta,* and the Bushmaster known from the Osa Peninsula of Costa Rica and from adjacent western Panama, *L. melanocephala.* Greene and Santana (1983) monitored a single individual long enough to record the time between successive captures of prey items. From this observation they estimated that only six meals are needed each year for the maintenance of an individual snake.

HOG-NOSED VIPER   *Porthidium nasutum*

Pls. 139, 140

OTHER COMMON NAMES: Tamaga, Jumping Viper, Hog-nosed Pitviper, Rainforest Hog-nosed Pitviper

SIZE: Largest male 410 mm TL (n = 3); largest female 405 mm TL (n = 5).

DISTRIBUTION: Caribbean slopes from Chiapas, Mexico, to northern Colombia; Pacific slopes from the Osa Peninsula, Costa Rica, to Ecuador.

IDENTIFYING FEATURES: This small, pudgy, terrestrial snake is probably the second most common pitviper at La Selva. It is easily distinguished from others by the presence of an enlarged, upturned rostral scale at the tip of the snout, hence the English common name, Hog-nosed Viper. The dorsal ground color is light grayish brown. A series of squareish, dark brown markings (each three to five scale rows wide) is found along the middorsum; these markings may become wider laterally, creating a series of faded, hourglass shapes. The top of the head is cinnamon brown in front of the eyes, shading to dark grayish brown behind the eyes. A dark stripe extends from the eye to the angle of the jaw; the sides of the face between this stripe and the top of the head are cinnamon brown. No known mimic of this species exists at La Selva.

SEXUAL DIMORPHISM: Elsewhere, the female is larger than the male (Lee 1996). However, this has not been determined for specimens at La Selva, and our scant data do not appear to conform to this pattern.

HABITAT: This species is most often encountered on leaf litter on trails in primary forest. Because its color pattern matches this background exceptionally well, more individuals must escape detection than are observed.

DIET: Prey items for the Hog-nosed Viper at La Selva include the Brilliant Forest Frog *(Rana warszewitschii)* (documented in 2000 in a photo by Dan Warner), the Wedge-billed Woodcreeper *(Glyphorhynchus spirurus),* and the Spiny Pocket Mouse *(Heteromys desmarestianus)* (Greene 1997). Additional prey items known from elsewhere within the range of this species include frogs of the genus *Eleutherodactylus,* lizards of the genus *Ameiva,* and small rodents (Alvarez del Toro 1983; photos in Greene 1997). The juvenile has a bright yellow tail tip that may be used to lure prey (Neill 1960).

REPRODUCTION: This species is viviparous. Litter sizes of 14 and 18 are known from elsewhere in Costa Rica (Picado 1976; Porras et al. 1981).

REMARKS: Although not particularly aggressive, this snake strikes in a peculiar fashion. Because it is so short and squat, the body seems to straighten completely, creating the appearance that the animal leaps forward. This is the source of a second English common name, the Jumping Viper. Four of six recent snakebite incidents at La Selva involved this snake (Hardy 1994).

Plate 77. Common
Snapping Turtle
(male).

Plate 78.
Narrow-bridged
Mud Turtle (male).

Plate 79.
White-lipped
Mud Turtle.

Plate 80.
White-lipped
Mud Turtle.

Plate 81.
White-lipped
Mud Turtle.

Plate 82. Brown
Wood Turtle.

Plate 83. Brown
Wood Turtle.

Plate 84. Black
Wood Turtle
(female).

Plate 85. Black Wood Turtle (male).

Plate 86. Black Wood Turtle (male).

Plate 87. Green Basilisk (male).

Plate 88. Striped Basilisk (female).

Plate 89. Striped Basilisk (male). Photo taken in Miami, Florida.

Plate 90. Casque-headed Lizard.

Plate 91. Green Iguana (male).

Plate 92. Green Iguana (juvenile male).

Plate 93. Green Tree Anole (male).

Plate 94. Green Tree Anole (female).

Plate 95. Pug-nosed Anole (male).

Plate 96. Pug-nosed Anole (juvenile female). Photo taken in Corcovado National Park, Costa Rica.

Plate 97. Carpenter's Anole (male).

Plate 98. Carpenter's Anole (juvenile).

Plate 99. Ground Anole (male).

Plate 100. Lemur Anole (female).

Plate 101. Lemur Anole (male).

Plate 102. Lemur Anole (female).

Plate 103. Slender Anole (male).

Plate 104. Slender Anole (male).

Plate 105. Stream Anole (male).

Plate 106. Lichen Anole (male).

Plate 107. Lichen Anole (juvenile).

Plate 108. Neotropical Chameleon (male). Photo taken in Guapiles, Costa Rica.

Plate 109.
Neotropical
Chameleon.

Plate 110. Yellow-
headed Gecko
(male).

Plate 111. Yellow-
headed Gecko
(female).

Plate 112.
Litter Gecko.

Plate 113. Yellow-tailed Dwarf Gecko (male).

Plate 114. Yellow-tailed Dwarf Gecko (female).

Plate 115. Yellow-tailed Dwarf Gecko (juvenile).

Plate 116. Spotted Dwarf Gecko.

Plate 117.
Turnip-tailed
Gecko.

Plate 118.
Turnip-tailed
Gecko.

Plate 119. Tropical
Night Lizard.

Plate 120. Tropical
Night Lizard
(juvenile).

Plate 121. Bronze-backed Climbing Skink. Photo taken in Santa Rosa National Park, Costa Rica.

Plate 122. Litter Skink.

Plate 123. Central American Whiptail.

Plate 124. Central American Whiptail (juvenile).

Plate 125. Central American Whiptail (male).

Plate 126. Rainforest Celestus.

Plate 127. Rainforest Celestus.

Plate 128. Talamancan Galliwasp.

Plate 129. Coral-mimic Galliwasp.

Plate 130. Boa Constrictor (juvenile). Photo taken in Volcán Cacao, Costa Rica.

Plate 131. Annulated Tree Boa.

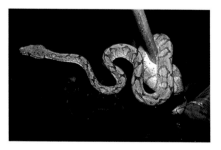

Plate 132. Annulated Tree Boa.

Plate 133.
Central American
Dwarf Boa.

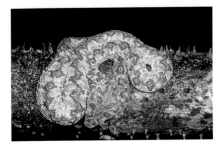

Plate 134.
Eyelash Viper.

Plate 135.
Eyelash Viper.

Plate 136.
Terciopelo.

Plate 137.
Bushmaster.

Plate 138.
Bushmaster.

Plate 139.
Hog-nosed Viper.

Plate 140.
Hog-nosed Viper.

Plate 141.
Ridge-nosed Snake.

Plate 142.
Ebony Keelback
(adult).

Plate 143. Ebony
Keelback (juvenile).

Plate 144. Clelia
(adult).

Plate 145.
Clelia (juvenile).

Plate 146. Brown
Debris Snake.

Plate 147. Brown
Debris Snake.

Plate 148. Brown
Forest Racer.

Plate 149.
Lowland Forest
Racer.

Plate 150.
Bicolored
Snaileater.

Plate 151.
Black-tailed Cribo.
Photo taken in
Pitilla, Costa Rica.

Plate 152.
Speckled Racer.

Plate 153.
Lower-montane
Green Racer.

Plate 154.
White-headed
Snake.

Plate 155.
Central American
Coralsnake Mimic
(juvenile).

Plate 156.
Hoffmann's
Earthsnake.

Plate 157. Tropical Seep Snake.

Plate 158. Brown Blunt-headed Vinesnake.

Plate 159. Yellow Blunt-headed Vinesnake.

Plate 160. Yellow Blunt-headed Vinesnake.

Plate 161.
Tropical Milksnake.

Plate 162.
Southern
Cat-eyed Snake.

Plate 163.
Northern
Cat-eyed Snake
(adult).

Plate 164.
Northern Cat-eyed
Snake (juvenile).

Plate 165.
Green Parrotsnake.

Plate 166.
Green Parrotsnake.

Plate 167.
Satiny Parrotsnake
(adult).

Plate 168.
Satiny Parrotsnake
(juvenile).

Plate 169. Striped Parrotsnake.

Plate 170. Fire-bellied Snake. Photo taken in Volcán Cacao, Costa Rica.

Plate 171. Fire-bellied Snake. Photo taken in Monteverde, Costa Rica.

Plate 172. Salmon-bellied Racer (adult). Photo taken in Dominical, Costa Rica.

Plate 173.
Salmon-bellied
Racer (juvenile).

Plate 174.
Spotted
Woodsnake.

Plate 175.
Spotted
Woodsnake.

Plate 176.
Red Coffeesnake.

Plate 177. Red Coffeesnake.

Plate 178. Rugose Littersnake.

Plate 179. Brown Vinesnake.

Plate 180. Short-nosed Vinesnake.

Plate 181.
Green Vinesnake.

Plate 182.
Calico Snake
(adult).

Plate 183. Calico
Snake (juvenile).

Plate 184.
Bird-eating Snake.

Plate 185. Bird-eating Snake.

Plate 186. Pink-bellied Littersnake.

Plate 187. Skinkeater (juvenile).

Plate 188. Skinkeater (adult).

Plate 189.
Ringed Slugeater.

Plate 190. Lichen-
colored Slugeater.

Plate 191.
Cloudy Slugeater.

Plate 192. Tiger
Ratsnake. Photo
taken in Corcovado
National Park,
Costa Rica.

Plate 193.
Reticulated
Crowned Snake.

Plate 194.
Orange-bellied
Crowned Snake.

Plate 195.
Tricolored Crowned
Snake (juvenile).
Photo taken in
Darién, Panama.

Plate 196.
Orange-bellied
Swamp Snake.
Photo taken in San
Blas, Panama.

Plate 197. Faded Dwarf Snake.

Plate 198. Long-tailed Littersnake (juvenile).

Plate 199. Halloween Snake.

Plate 200. Orange-bellied Littersnake.

Plate 201. False Terciopelo.

Plate 202. Allen's Coralsnake. Photo taken in Arenal, Costa Rica.

Plate 203. Many-banded Coralsnake.

Plate 204. Central American Coralsnake.

Plate 205. Spectacled Caiman (juvenile).

Plate 206. Spectacled Caiman (juvenile).

Plate 207. American Crocodile. Photo taken in Palo Verde, Costa Rica.

Colubridae

This is a "garbage can" family because elapids are likely to be derived from within this group. Thus, some members of this family are more closely related to elapids than they are to other colubrids. For this reason, there is no convenient field characteristic for the family other than to eliminate the possibility that the subject in question is a member of any other family. Nevertheless, this is an extremely successful radiation of squamates. This family has over 1,700 named species, making it the most species-rich family of squamates. These snakes are found on all continents except Antarctica and are found in all habitats. There are 47 species of colubrids known from La Selva. Four additional species are known from elsewhere in the Caribbean lowland wet forests of Costa Rica.

Key to the Colubridae at La Selva

1a   Color pattern of bicolored or tricolored rings or bands (fig. 18a, d; these may be faded in large individuals and may be as narrow as one scale row wide) of red (or pink or orange), black, and white or yellow 2

1b   Color pattern variable but not including brightly colored (red, black, yellow, or white) rings or bands 9

    2a   Body bicolored throughout length (red, orange, or white, and black; select this choice for forms in which the dorsal half of the light band is orange and the ventral half is white or yellowish) 3

    2b   Body tricolored (red or orange, black, and yellow or white) 5

3a   Color pattern of bands (i.e., black color not extending onto venter); posterior border of black skullcap U-shaped; anterior-most light bands yellow to white, shading to orange posteriorly . . . . *Oxyrhopus petolarius*

3b   Color pattern of rings (i.e., black color extending onto venter); posterior border of black skullcap straight; light bands including orange or white throughout length of body 4

    4a   Snout mottled with orange, black, and white . . . . *Dipsas bicolor*

    4b   Snout solid black. . . . . *Urotheca euryzona*

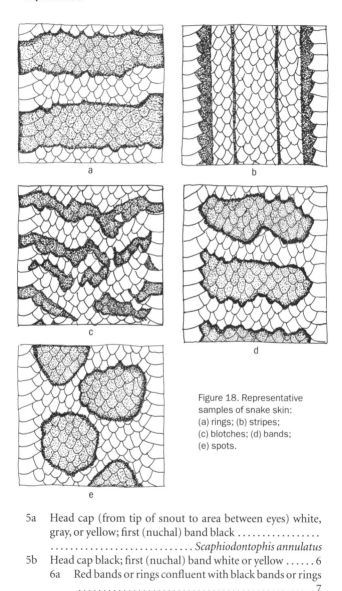

Figure 18. Representative samples of snake skin: (a) rings; (b) stripes; (c) blotches; (d) bands; (e) spots.

5a Head cap (from tip of snout to area between eyes) white, gray, or yellow; first (nuchal) band black . . . .
. . . . *Scaphiodontophis annulatus*
5b Head cap black; first (nuchal) band white or yellow 6
6a Red bands or rings confluent with black bands or rings
. . . . 7

6b   Red bands or rings confluent with yellow or white bands or rings 8

7a   Body with bands (i.e., black color not extending onto venter); light and dark bands often offset at middorsum . . . . . .
. . . . *Tantilla supracincta*

7b   Body with rings (i.e., black color extending onto venter); light and dark rings typically not offset at middorsum . . . . .
. . . . *Lampropeltis triangulum*

8a   Body with bands (i.e., black color not extending onto venter); body bicolored anteriorly, tricolored posteriorly . . . . *Oxyrhopus petolarius*

8b   Body with rings (i.e., black color extending onto venter); body tricolored throughout (excluding tail) . . . . .
. . . . *Erythrolamprus mimus*

9a   Basic body color green or olive on at least the anterior half of body, or body with olive and brown bands 10

9b   Basic body color not including green on anterior half of body 18

10a   Anterior half of body olive green; posterior half black or dark gray . . . . *Drymarchon corais*

10b   Color pattern more or less consistent throughout length of body 11

11a   Body more or less uniform leaf green (sometimes with thin, dorsolateral, dark stripes or a broad, bronze, middorsal stripe) 12

11b   Body with banded or lichenose pattern involving red, yellow, and/or brown 17

12a   Ventral-most dorsal scales light (white, yellow, or bronze), confluent with light coloration of ventral scutes or forming a light, lateral stripe 13

12b   All dorsal scales green; no light, lateral stripe 14

13a   Lateral stripe bordered below by green, separating light stripe from belly coloration; no middorsal, bronze stripe . .
. . . . *Oxybelis fulgidus*

13b   Lateral light coloration confluent with light belly coloration; wide, middorsal, bronze stripe separated from light belly color by dark green or bluish lateral coloration . . . . . .
. . . . *Leptophis nebulosus*

14a   Two to five thin, black stripes along vertebral and paravertebral scale keels (fig. 18b) 15

14b   No thin, black vertebral or paravertebral stripes . . . 16

15a A loreal scale present (fig. 19a) .... *Leptophis depressirostris*
15b No loreal scale (fig. 19b).............. *Leptophis ahaetulla*

   16a Anterior tips of dorsal scales yellow.................
      .... *Drymobius melanotropis*
   16b Dorsal scales uniform green ..... *Oxybelis brevirostris*

17a Ventrolateral area patterned differently than middorsum
(ventrolateral area covered by white blotches separated by
light brown; middorsum with olive green ground color and
blotches of dark and light brown); upper lip not yellow....
.... *Sibon longifrenis*
17b Ventrolateral area colored like dorsum (body banded or
with mottled patches of brick red and olive green); upper lip
yellow or white .... *Pseustes poecilonotus*

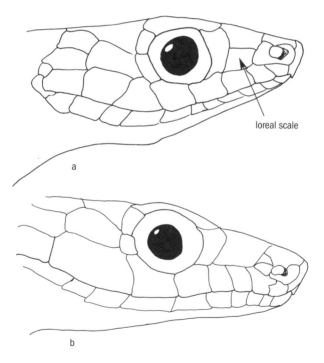

loreal scale

a

b

Figure 19. Lateral view of head: (a) *Leptophis depressirostris* (with loreal scale); (b) *Leptophis ahaetulla* (without loreal scale).

18a   Basic body color red (including brick red)......... 19

18b   Basic body color not red......................... 20

19a Two light nuchal bands (fig. 20a); scales keeled ....
.... *Ninia sebae*

19b One light nuchal band (fig. 20b); scales smooth...........
.... *Clelia clelia*

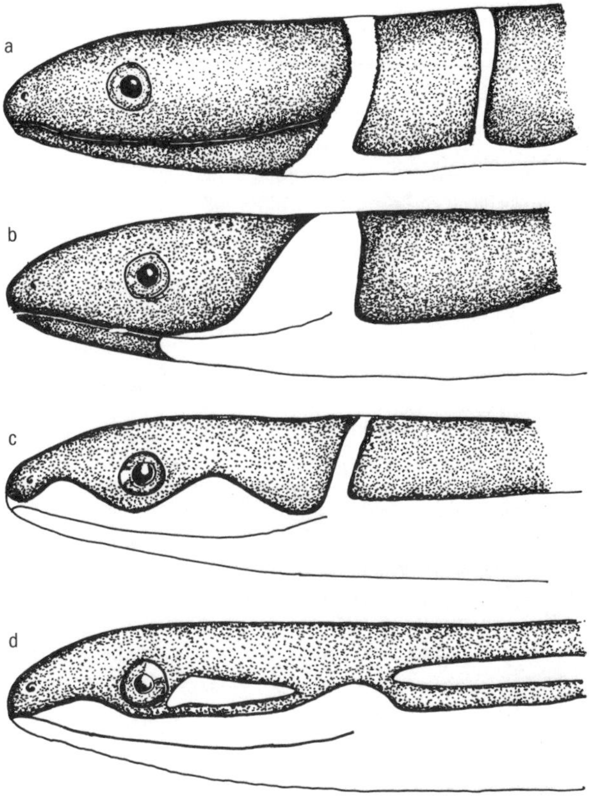

Figure 20. Stylized lateral view of a snake head: (a) two complete light nuchal collars *(Ninia sebae)*; (b) one complete light nuchal collar (e.g., juvenile *Clelia clelia*); (c) one interrupted light nuchal collar *(Tantilla reticulata)*; (d) light nuchal (cheek) patch *(Tantilla ruficeps)*.

20a Top of head covered by large, symmetrical plates. . . 21
20b Top of head covered by small, irregular scales or a mixture of small and large scales . . . . *Nothopsis rugosus*
21a Basic body color black, gray, or speckled with light and dark
. . . . 22
21b Basic body color brown 32
22a Body uniform black with white upper lip and belly. . .
. . . . *Chironius grandisquamis*
22b Body variously colored but not as above 23
23a Body black with thin, irregular, yellow bands; head often boldly marked with black and yellow bands; belly yellow . .
. . . . *Spilotes pullatus*
23b Body variously colored but not as above 24
24a Top of head bone white. 25
24b Top of head variously colored but not bone white. . 26
25a Body black with very thin, white bands . . . .
. . . . *Scaphiodontophis annulatus*
25b Body uniform black to gray . . . . *Enulius sclateri*
26a Body mostly uniform gray (may have dark or light nuchal collar and/or faint orange spots) 27
26b Body patterned with distinct small spots, large blotches, or bands involving dark gray, black, and/or white 29
27a Body large, thick, and muscular for constriction; scales with iridescent sheen . . . . *Clelia clelia*
27b Body slight and not designed for constriction; scales without pronounced iridescent sheen. 28
28a Venter white or light gray, changing abruptly to darker dorsal coloration . . . . *Geophis hoffmanni*
28b Venter gray, changing gradually to darker gray dorsal coloration . . . . *Amastridium veliferum*
29a Dorsum mottled with light and dark gray blotches (fig. 18c)
. . . . 30
29b Dorsum with dark and light bands or with dark and light speckles. 31
30a Body triangular in cross section with a noticeable middorsal ridge; scales strongly keeled. . . . *Ninia maculata*
30b Midbody round in cross section (no ridge); scales smooth. . . . . *Sibon nebulatus*
31a Dorsum banded with black and light; interscale epidermis light blue and red creating a coral-mimicking pattern when lung inflated. . . . . *Liophis epinephalus*

31b Dorsum speckled with light and dark (salt-and-pepper), often with a bluish cast; interscale epidermis not light blue or red . . . . *Drymobius margaritiferus*

    32a Dorsum uniform brown or with distinct light or dark stripes (fig. 18b) 33

    32b Dorsum blotched, cross-banded, or spotted (fig. 18c, d, e) 42

33a A light (white, yellow, or tan) nuchal collar or a light, lateral patch on each side of body in nuchal region (fig. 20b, c, d) 34

33b No light nuchal collar or patches 37

    34a An entire, tan nuchal collar (fig. 20b); venter white with black spot at lateral margin of each scute 35

    34b A dissected, white nuchal collar or light, lateral patches in nuchal region (fig. 20c, d); venter yellow or orange pink 36

35a A pair of distinct, light, lateral lines along length of body . . . . *Urotheca decipiens*

35b Body more or less uniform in coloration . . . . *Trimetopon pliolepis*

    36a A well-developed nuchal collar interrupted only at middorsum (fig. 20c); upper lip dark except for light patches immediately anterior and posterior to eye; venter yellow . . . . *Tantilla reticulata*

    36b A pair of lateral, light nuchal patches, widely separated at middorsum (fig. 20d); upper lip white except for a dark subocular spot; venter orange or pink . . . . *Tantilla ruficeps*

37a Body with light stripes along entire length. . . . . *Urotheca guentheri*

37b Body uniform brown, or if light stripes present, these fading noticeably past anterior fourth of body 38

    38a Dorsum uniform tan or with small, dark spots . . . . *Hydromorphus concolor*

    38b Dorsum with faint, dorsolateral, light stripes 39

39a Venter salmon or red. 40

39b Venter white or light yellow. 41

    40a Bold white spots and/or stripes behind eye and on nape of neck, these confluent with a white, dorsolateral stripe that fades dramatically by posterior three-fourths of body; venter tomato red . . . . *Rhadinaea decorata*

40b No bold, white spots or stripes behind eye or on nape of neck; dorsolateral stripe faint throughout length of body; venter salmon . . . . . . *Mastigodryas melanolomus*

41a Venter immaculate; upper lip uniform yellow or white . . . .
. . . . *Dendrophidion percarinatum*

41b Venter with small, lateral, black spots; upper lip same color as rest of head (brown) . . . . *Coniophanes fissidens*

42a Top of head and neck dominated by a distinct, dark, heart-shaped blotch bordered by white or light yellow
. . . . *Xenodon rabdocephalus*

42b No distinct, dark blotch on top of head and neck, or if blotch present, then not heart-shaped 44

43a Dorsum with dark chocolate brown bars bordered for dorsal third by olive green and for ventral two thirds by white
. . . . *Sibon annulatus*

43b Dorsum variously colored but not as above 44

44a Head elongate and pointed; dark stripe from tip of snout through eye to angle of jaw; lining of mouth dark purple to black . . . . *Oxybelis aeneus*

44b Head not unusually elongate and pointed; no black eye mask; lining of mouth not dark purple to black. . .
. . . . 45

45a Head noticeably blunt; body unusually thin 46

45b Head normal in shape; body not unusually thin 47

46a Dorsum light grayish tan with bold, diamond-shaped, dark brown markings down midline . . . .
. . . . *Imantodes cenchoa*

46b Dorsum without bold, dark, diamond-shaped markings . . . . *Imantodes inornatus*

47a A ventrolateral, light stripe one scale row wide separated from white, immaculate belly scutes by two rows of dark scales (ventral-most dorsal scales); a series of irregular spots, generally in paravertebral pairs and larger for anterior half (five to six scales per spot) than posterior half (one to two scales per spot) of body. . . *Tretanorhinus nigroluteus*

47b No ventrolateral, light stripe; dorsum without spots 48

48a Anterior portion of dorsum with light and dark cross bands. 49

48b Anterior portion of dorsum with dark blotches or with a zigzag marking 50

49a Vivid, dark-edged, light cross bands on neck; interscale epidermis blue . . . . *Dendrophidion vinitor*

49b Narrow, dark-edged, tan bands anteriorly and dark, longitudinal stripes posteriorly; interscale epidermis not blue
.... *Dendrophidion percarinatum*

50a Venter pinkish to salmon, chin dark brown with light spots and blotches; upper and lower labials with distinct light and dark bars; dorsum with a series of squareish to rounded, dark, middorsal blotches on a light brown background, with a series of dark, lateral blotches offset from middorsal ones ....
.... *Mastigodryas melanolomus*

50b Venter and chin white; upper and lower labials not barred with light and dark; dorsum with rounded blotches or a zigzag pattern 51

51a Body reddish brown with dark brown blotches along middorsum that only occasionally meet to form a zigzag pattern; a bold, white nuchal collar in the juvenile............
.... *Leptodeira septentrionalis*

51b Body brown but not reddish; a zigzag-shaped, dark marking down middle of back .... *Leptodeira annulata*

RIDGE-NOSED SNAKE   *Amastridium veliferum*
Pl. 141
OTHER COMMON NAMES: Rustyhead Snake
SIZE: Largest male 482 mm TL (n = 16);
largest female 376 mm TL (n = 9).
DISTRIBUTION: Caribbean slopes from Nicaragua to Panama; Pacific slopes from the Osa Peninsula, Costa Rica, to Panama.

IDENTIFYING FEATURES: This is a small leaf litter snake. The head is dark reddish brown with very thin, gray stripes. The nuchal region has an interrupted, light gray collar, which appears as two light patches on each side of the face and neck, disrupted by a lanceolate extension of the dark dorsal color. The dorsal ground color is dark smoky gray with light brown punctations that are bordered by black. Middorsal and paravertebral, black stripes are present but difficult to see because they are so thin. The ventral surface of the head is brown with white and gray flecks. The anterior third of the venter is gray with brown flecks, shading to uniform gray posteriorly. Only the White-headed Snake *(Enulius sclateri)* and Hoffmann's Earthsnake *(Geophis hoffmanni)* have a dorsal color pattern similar to the Ridge-nosed Snake. However, the White-headed Snake has a light tan nuchal collar (at most light patches in the Ridge-nosed Snake), and Hoffmann's Earth-snake has an abrupt change from light ventral color to dark dorsal color (gradual change in the Ridge-nosed Snake).

SEXUAL DIMORPHISM: No external features distinguish the two sexes. The adult male and female have approximately the same total lengths at La Selva (means of 350 mm and 330 mm, respectively).

HABITAT: We marked and released 25 individuals of this species from 1982 to 1984. Apparently, this is the largest series of specimens ever observed (see Lee 1996). All were collected during the day when they were discovered crossing trails, principally in primary forest.

DIET: This snake is rear-fanged and consumes frogs, mostly of the genus *Eleutherodactylus* (Blaney and Blaney 1978; Martin 1955).

REPRODUCTION: We palpated one female on May 4, 1983, that appeared to have three oviductal eggs. From this, we believe the species to be oviparous, but nothing further is known regarding reproduction in this snake (Lee 1996).

REMARKS: Savage (2002) places this genus, long considered an enigma to systematic herpetologists, along with the genera *Conophis, Crisantophis, Enulius,* and *Nothopsis,* into an informal group *(Amastridines).*

EBONY KEELBACK
Pls. 142, 143

Chironius grandisquamis

OTHER COMMON NAMES: Zopilota, Ebony Keelback Snake

SIZE: Largest male 2,210 mm TL (n = 11); largest female 2,010 mm TL (n = 5).

DISTRIBUTION: Caribbean slopes from Honduras to Panama; Pacific slopes from central Costa Rica to Ecuador.

IDENTIFYING FEATURES: This is one of two common, large, dark, diurnal snakes at La Selva. The adult and juvenile differ in dorsal coloration. The adult is more or less uniform black. The head is dark with an abrupt change to bone white on the ventral portion of the supralabials (upper lip scales). This gives the animal a white-lipped appearance. The venter is white anteriorly, fading to cream or light gray posteriorly. The juvenile possesses a banded dorsum of black and dark brown bars; the bars are sometimes separated by thin, light cream bands. In both the adult and juvenile, the dorsal scales are extremely large and easily visible from a safe distance (1 to 2 m). The adult might be confused with the adult of the Clelia *(Clelia clelia)* or the Tiger Ratsnake *(Spilotes pullatus)*. The Clelia tends to be lighter in color (gray rather than black), bulkier in body shape, and less aggressive than the Ebony Keelback; the Tiger Ratsnake differs in having bold, yellow markings on the head and neck. The juvenile Ebony Keelback is most similar to the juvenile Salmon-bellied Racer *(Mastigodryas melanolomus)*; the latter can be distinguished from the former by the presence of a reddish to orangish belly.

SEXUAL DIMORPHISM: No external features distinguish the two sexes. Adult males in our samples from La Selva averaged slightly longer total length (2,000 mm) than females (1,760 mm).

HABITAT: This species is an active, diurnal resident of primary and secondary forests. It is usually encountered crossing trails but climbs readily and may coil in trees above streams, where it presumably roosts for the night.

DIET: The Ebony Keelback eats anurans and is unusual among La Selva's frog-eating snakes in consuming the Smoky Jungle Frog *(Leptodactylus pentadactylus)*, a prey item with potent skin toxins. We have also observed it to eat adult the Common Mexican Treefrog *(Smilisca baudinii)* and the Reticulated Sheepfrog *(Gastrophryne pictiventris)*. Dixon et al. (1993) reports the diet to include salamanders and leptodactylids, and Scott (1969) lists birds, mice, lizards, and frogs as prey items.

REPRODUCTION: This snake is oviparous and may produce up to 15 eggs (Dixon et al. 1993).

REMARKS: This species is aggressive and gives threat displays and strikes when disturbed. In the threat display, the snake gapes its jaws, inflates its lung (exposing the white skin between the large scales), laterally compresses itself in the neck region, and hisses.

CLELIA

Clelia clelia

Pls. 144, 145

OTHER COMMON NAMES: Zopilota, Mussurana

SIZE: Largest male 2,210 mm TL (n = 3); largest female 2,179 mm TL (n = 3).

DISTRIBUTION: Caribbean slopes from Veracruz, Mexico, to Uruguay and Argentina; Pacific slopes from Guerrero, Mexico, to Peru.

IDENTIFYING FEATURES: This large snake is striking for its dramatic color change from juvenile to adult. The adult is a uniform deep steel gray. The scales are smooth, giving the animal an iridescent sheen. The dorsal color shades gradually to off-white on the venter, which is unmarked except for scattered, gray spots on the subcaudals. The juvenile is uniform bright red except for the anterior end, which is marked with bold black and light bands. The head cap of the juvenile is black, bordered by a white or light yellow nuchal collar. The collar is bordered posteriorly by a black band that changes relatively abruptly to red. The adult of this species is not easily mistaken for any other snake at La Selva except the Ebony Keelback *(Chironius grandisquamis);* these two can be distinguished by the generally darker coloration, noticeably enlarged dorsal scales, and distinctive white lip (supralabials) of the Ebony Keelback and the heavyset body and sluggish disposition of the Clelia. The juvenile Clelia has a color pattern that is similar to the Red Coffeesnake *(Ninia sebae).* However, the Red Coffeesnake has keeled scales, paired anterior, light bands, and a dull red coloring.

SEXUAL DIMORPHISM: No external features distinguish the two sexes.

HABITAT: This species is found in primary forest, where it may be active day or night. It is typically found on the ground.

DIET: The Clelia is a powerful constrictor that feeds on snakes, mammals, and occasionally lizards (Duellman 1978; Scott 1983b; Sexton and Heatwole 1965). The species is relatively unaffected by pitviper venom. It is the largest of La Selva's rear-fanged colubrids; the rear fangs are associated with venom glands that produce toxins known to kill mammalian prey. Photographic records from La Selva document the Nine-banded Armadillo *(Dasypus novemcinctus)* and the Terciopelo *(Bothrops asper)* as prey.

REPRODUCTION: This snake lays 10 to 20 eggs (Duellman 1963; Martinez and Cerdas 1986); nothing else is known regarding patterns of reproduction. Leenders (2001) reports an individual from Costa Rica that produced a clutch of 10 large eggs: they averaged 61 mm in length and hatched in four months.

BROWN DEBRIS SNAKE   *Coniophanes fissidens*

Pls. 146, 147

OTHER COMMON NAMES: Yellowbelly Snake, White-lipped Spotbelly Snake

SIZE: Largest male 524 mm TL (n = 12); largest female 513 mm TL (n = 15).

DISTRIBUTION: Caribbean slopes from southern San Luis Potosí, Mexico, to northern Colombia; Pacific slopes from Michoacán, Mexico, to Ecuador.

IDENTIFYING FEATURES: This species is a small, stout ground snake. The dorsal ground color is reddish or cinnamon brown. Each middorsal scale has a small, black punctation, creating a thin, middorsal, dark stripe. The ventral-most four scale rows form a faint coppery brown stripe from the neck to the level of the vent. The venter is white, light yellow, or peach shading to orange laterally. Each ventral scute has several small, dark punctations. The head is brown with a thin, light stripe (bordered above by dark) from the back of the eye to the angle of the jaw. A pair of white spots or short stripes is found from the angle of the jaw to the neck. The Brown Debris Snake is most similar to the Pink-bellied Littersnake *(Rhadinaea decorata)* in dorsal color pattern. The two can be distinguished by the thinner body and immaculate venter of the Pink-bellied Littersnake.

SEXUAL DIMORPHISM: We know of no external features that distinguish the two sexes. At La Selva, the adult male and female have similar mean total lengths (451 mm for the male; 430 mm for the female).

HABITAT: This snake is found most frequently crossing trails by day in primary and secondary forest. It is a member of an assemblage of small snakes that live in the leaf litter.

DIET: This snake consumes frogs, lizards, salamanders, and invertebrates (Myers 1969a; Seib 1985a).

REPRODUCTION: This species is oviparous with a modal clutch size of three eggs; eggs may be laid during the dry season and hatch during the wet season (Zug et al. 1979). We observed two apparently gravid females, one in June, 1983, and one on December 30, 1984, that contained three and four oviductal eggs, respectively.

REMARKS: Bites to humans by this snake are known to cause swelling, numbness, and/or pain. Fortunately, it rarely bites, even when handled. The species is among those small leaf litter colubrids that have proportionately long tails that are easily lost (Guyer and Donnelly 1990).

BROWN FOREST RACER   *Dendrophidion percarinatum*
Pl. 148

SIZE: Largest male 1,055 mm TL (n = 8); largest female 888 mm TL (n = 1).

DISTRIBUTION: Caribbean slopes from northern Honduras to Panama; Pacific slopes from the Osa Peninsula, Costa Rica, to Ecuador.

IDENTIFYING FEATURES: The adult of this medium-sized species appears to have a uniform gray brown dorsum, but close inspection, especially of the posterior third of the animal, reveals a pair of faint, dorsolateral, light stripes on each side, the second stripe formed on the ventral-most dorsal scales. The venter is light yellow anteriorly shading to immaculate white posteriorly. The head is gray brown with yellow or white supralabials. Smaller individuals have a dorsal coloration that consists of thin, tan bands bordered by dark. Posteriorly this color pattern fades to uniform gray brown with paired, lateral, light stripes. The Brown Forest Racer can be distinguished from the Lowland Forest Racer *(D. vinitor)* by its faint stripes rather than thin bands and by the lack of blue between the anterior scales (see the species account for the Lowland Forest Racer). The Brown Forest Racer is also similar in shape and habits to the Salmon-bellied Racer *(Mastigodryas melanolomus),* from which it can be distinguished by its yellow to white, rather than a salmon-colored, venter.

SEXUAL DIMORPHISM: No external features distinguish the two sexes.

HABITAT: All individuals that we have observed were captured in abandoned cacao *(Theobroma cacao)* plantations. This species is less common at La Selva than its congener the Lowland Forest Racer.

DIET: The species presumably is an active diurnal predator of frogs.

REPRODUCTION: Presumably this species lays eggs.

REMARKS: This snake has an unusually long tail for a colubrid, and the tail is easily lost (Guyer and Donnelly 1990).

LOWLAND FOREST RACER
Dendrophidion vinitor

Pl. 149

OTHER COMMON NAMES: Barred Forest Racer
SIZE: Largest male 1,025 mm TL (n = 14);
largest female 1,100 mm TL (n = 14).
DISTRIBUTION: Caribbean slopes from southern
Veracruz, Mexico, to Panama; Pacific slopes from cen-
tral Costa Rica to Panama.

IDENTIFYING FEATURES: This midsized species is one of the more common racerlike snakes at La Selva. The dorsum is gray brown for the anterior fourth to third, shading to brown posteriorly. The anterior region is crossed by wide, dark bands (three to five scale rows wide), the anterior-most of which is rusty red, whereas the rest are gray brown. Each dark band is followed by a thin, light, grayish tan band (one scale row wide) bordered by dark. The skin between the scales in the anterior region is light blue, creating a bluish tint to the light bands under normal conditions and a blue band when the lung is expanded. In the posterior region, a heavily dissected, middorsal, tan stripe is found; the dissection of this stripe is caused by thin, dark bands that shade to gray laterally. The venter is immaculate white to light yellow. The head is gray brown, and the supralabials are brownish anteriorly and white to light yellow posteriorly. Two species at La Selva are likely to be confused with this one. However, one of these, the Brown Forest Racer *(D. percarinatum)*, is uniform in color anteriorly and faintly striped posteriorly, and the other, the Salmon-bellied Racer *(Mastigodryas melanolomus)*, is salmon colored on the venter.

SEXUAL DIMORPHISM: We know of no external features that distinguish the two sexes; the male is slightly shorter in mean total length than the female (810 mm and 920 mm, respectively).

HABITAT: This species is typically observed crossing trails in primary and secondary forest. It is wary, has particularly good vision, and usually races away when approached. It is also an adept climber. The species can be found at night coiled in understory shrubs and trees.

DIET: Frogs are the primary diet item; at La Selva, this snake consumes Bransford's Litterfrog *(Eleutherodactylus bransfordii)* and the Harlequin Treefrog *(Hyla ebraccata)*.

REMARKS: This species has an unusually long tail that is often broken (Guyer and Donnelly 1990).

BICOLORED SNAILEATER

Dipsas bicolor

Pl. 150

SIZE: We have no measurements of this species from La Selva; elsewhere in Costa Rica this species reaches sizes of 586 mm TL (Campbell and Lamar 1989), and La Selva individuals are likely similar in size.

DISTRIBUTION: Caribbean slopes from southeastern Honduras to Costa Rica.

IDENTIFYING FEATURES: This arboreal snake is characterized by a pattern of light (orangish) and dark (black) rings. The light rings are seven to eight scale rows wide and number about 25 along the length of the body. These rings fade to white ventrally. Additionally, the light rings fade along the anterior and posterior margins, but not enough to create a tricolored pattern. The dark rings can be up to about 20 scale rows wide at midbody but are closer in size to the light bands near the head and the tail. The head is mottled with orange, black, and white, especially along the supralabials and on the snout, and the first body ring posterior to the head is a light one. This species is most similar to the Ringed Slugeater *(Sibon annulatus)* but differs from that species in having orangish light rings.

SEXUAL DIMORPHISM: No external features distinguish the two sexes.

HABITAT: This species is arboreal and nocturnal.

DIET: This species, like other members of the genus *Dipsas,* consumes gastropods, which it pulls from the shell with specialized elongate teeth on the lower jaw.

REMARKS: This species was not documented at La Selva until 1999, after which it has been recorded every year.

BLACK-TAILED CRIBO

Drymarchon corais

Pl. 151

OTHER COMMON NAMES: Zopilota, Indigo Snake, Central American Cribo

SIZE: We have no measurements of this species from La Selva; elsewhere this snake reaches sizes of ca. 2,950 mm TL (Wilson and Meyer 1985).

DISTRIBUTION: Caribbean slopes from southern Texas to Argentina; Pacific slopes from Sonora, Mexico, to Peru.

IDENTIFYING FEATURES: This large snake is olive green on the anterior half of the body, shading relatively abruptly to black or dark gray on the posterior half. The greenish region is checkered faintly with small, dark markings. The head is uniform olive green except for a broad, dark bar extending from below the eye to the upper lip. No other snake of similar size at La Selva is similar in coloration.

SEXUAL DIMORPHISM: Sometimes the adult male of this species develops faint keels on one to five middorsal scale rows starting about one-fourth the distance from the head and continuing posteriorly to the vent and sometimes onto the tail; the adult female has smooth scales (Layne and Steiner 1984). This characteristic needs to be confirmed for the tropical form of this snake.

HABITAT: This snake is diurnal and terrestrial. It occupies a variety of habitats, often in association with water (Lee 1996).

DIET: The species consumes a variety of vertebrates, but snakes tend to predominate in the diet (Duellman 1963; Greene 1975; Henderson and Hoevers 1977; Stuart 1948).

REPRODUCTION: This species is oviparous, but little else is known about patterns of reproduction for tropical populations.

REMARKS: A single photograph taken in the laboratory clearing in the 1970s is the only record of this species from La Selva. It is curious that such a large, diurnal predator was not reported before and has not been seen since. However, this species is known from the Atlantic versant elsewhere in Costa Rica, so it is not improbable that it might also inhabit La Selva.

SPECKLED RACER   *Drymobius margaritiferus*
Pl. 152
OTHER COMMON NAMES: Ranera
SIZE: Largest male 691 mm TL (n = 4);
largest female 942 mm TL (n = 9).
DISTRIBUTION: Caribbean slopes from southern
Texas to Colombia; Pacific slopes from Sonora,
Mexico, to Panama.
IDENTIFYING FEATURES: This is a handsome, midsized, salt-and-pepper-colored snake. The dorsal ground color is dark brown to black, but the median portion of each scale changes abruptly to a narrow ring of light blue and/or tan around a yellow center. The head is dark brown with yellow supralabials, the sutures of which are black. The head and body coloration are separated by a V-

shaped, yellow marking in the nuchal region. The venter is immaculate yellow, and the iris is rusty brown. The only species likely to be confused with the Speckled Racer at La Selva is the Lower-montane Green Racer *(D. melanotropis),* which is green in color.

SEXUAL DIMORPHISM: No external features distinguish the two sexes. At La Selva, the mean total length of the adult male is shorter than that of the female (600 mm and 720 mm, respectively).

HABITAT: This species is found primarily in secondary forest or disturbed, open areas, usually near water. It is diurnal and moves rapidly.

DIET: Frogs constitute most of the known diet records for this snake (Duellman 1963; Henderson and Hoevers 1977; Lee 1996; Ramirez et al. 1997; Seib 1984, 1985b; Smith 1943; Stuart 1935), although Scott (1969) lists lizards as well. At La Selva, this species is known to consume Bransford's Litterfrog *(Eleutherodactylus bransfordii),* Fitzinger's Rainfrog *(E. fitzingeri),* and the Reticulated Sheepfrog *(Gastrophryne pictiventris).*

REPRODUCTION: This snake lays eggs. Clutches of four and five eggs are known from Costa Rica, these laid in the dry season (Solórzano and Cerdas 1987).

LOWER-MONTANE GREEN RACER
Pl. 153

Drymobius melanotropis

OTHER COMMON NAMES: Ranera, Green Frog-eater

SIZE: Largest male 1,220 mm TL (n = 3); largest female 1,230 mm TL (n = 3).

DISTRIBUTION: Caribbean slopes from Honduras to Panama.

IDENTIFYING FEATURES: This snake is a rare, midsized forest racer at La Selva. The species has an olive green dorsal ground color shading to lime green laterally. Each dorsal scale is tipped with yellow; some are flecked with black in the region of the apical pits. The head is olive green washed with rusty brown, and the ventral scutes are white anteriorly, shading to bright yellow posteriorly. This species is most similar to the Bird-eating Snake *(Pseustes poecilonotus)* in appearance, but the latter has yellow supralabials and irregular reddish or yellowish banding. Additionally, the Lower-montane Green Racer tends to flee when disturbed, whereas the Bird-eating Snake often stands its ground.

SEXUAL DIMORPHISM: We know of no external features that distinguish the two sexes.

HABITAT: This species appears to be more abundant at midelevation sites in the Braulio Carrillo National Park, barely entering the hilly region at the southern end of La Selva. It is found in primary forest, often near water.

DIET: Frogs are presumed to be the primary diet item (Scott 1969).

REPRODUCTION: This species lays eggs. Two females from the Braulio Carrillo National Park extension contained two and three oviductal eggs.

WHITE-HEADED SNAKE   *Enulius sclateri*
Pl. 154

OTHER COMMON NAMES: Collared Snake
SIZE: Largest individual 404 mm TL (n = 3).
DISTRIBUTION: Caribbean slopes from
Nicaragua to Colombia; Pacific slopes from central Costa Rica to Panama.
IDENTIFYING FEATURES: This is a small species, most similar to the Ridge-nosed Snake *(Amastridium veliferum)* and Hoffmann's Earthsnake *(Geophis hoffmanni)* in general size and body coloration. The head typically is bone white, except for the very tip of the snout, which is dark gray, and the eye, which is jet black. Sometimes this species has a light gray rather than a white head, and a faded, tan nuchal collar may separate the head color from that of the rest of the body. The body is uniform dark gray to black for the entire length of the snake. The venter is white to light gray and can be nearly transparent. The White-headed Snake differs from the Ridge-nosed Snake and Hoffmann's Earthsnake in having a head that is distinctly lighter in color than the rest of the body.

SEXUAL DIMORPHISM: We know of no external features that distinguish the two sexes.

HABITAT: The species is rarely encountered, but it has been observed to be active both by day and by night, in or on leaf litter.

DIET: This snake has rear fangs and probably eats reptile eggs.

REPRODUCTION: This snake is presumed to be oviparous.

CENTRAL AMERICAN CORALSNAKE MIMIC

Erythrolamprus mimus

Pl. 155

OTHER COMMON NAMES: Coral Falsa, False Coralsnake

SIZE: Largest male 720 mm TL (n = 5); largest female 840 mm TL (n = 4).

DISTRIBUTION: Caribbean slopes from Honduras to Peru; Pacific slopes from the Osa Peninsula, Costa Rica, to Ecuador.

IDENTIFYING FEATURES: This midsized, rear-fanged snake mimics coralsnakes in color pattern. The head cap is black, extending from the tip of the snout to just posterior to the orbits (post-frontal suture). The black head cap is followed by a white ring that covers most of the parietals and posterior supralabials. A wide, black ring then covers the area posterior to the parietals, and this is followed by a thin, white ring. The dorsal pattern thereafter consists of a repetitive series of rings of the following colors: red (widest ring; seven to nine scale rows anteriorly, three to four posteriorly), white (narrowest ring; one to two scale rows), black (six to seven scale rows), and white again. The black rings usually are bisected laterally by a narrow, white band (one to two scale rows wide) that does not reach the middorsal portion of the black ring. Typically, there are 12 to 14 red rings. All red and white scales are tipped with black, muting their color. This pattern is a modified tricolored monad of the color regimes described by Savage and Slowinski (1992). Ventrally, the red and white rings change to dark gray. The dorsal rings often do not meet at the middorsal line, giving the dorsum a checkered appearance. Six other snake species at La Selva have a similar tricolored pattern. Two of these, the Tropical Milksnake *(Lampropeltis triangulum)* and the Tricolored Crowned Snake *(Tantilla supracincta),* differ in that the red color is bordered by black. All others have either more red rings or bands (more than 15; Allen's Coralsnake *[Micrurus alleni],* the Central American Coralsnake *[M. nigrocinctus],* and the Calico Snake *[Oxyrhopus petolarius])* or fewer (10; the Skinkeater *[Scaphiodontophis annulatus])* than the Central American Coralsnake Mimic.

SEXUAL DIMORPHISM: We know of no external features that distinguish the two sexes.

HABITAT: The species is found in primary and secondary forest, where it is active during the day and found on or in leaf litter.

DIET: The diet is composed principally of small snakes; in this respect the species mimics coralsnakes in diet as well as coloration. We palped an adult Red Coffeesnake *(Ninia sebae)* from the digestive tract of one individual and two snake eggs from a second individual.

REPRODUCTION: This snake is presumed to lay eggs.

REMARKS: Bites from this snake are known to cause pain and swelling in humans (Picado 1976). We observed one individual flatten its neck dorsoventrally, presumably in a threat display.

HOFFMANN'S EARTHSNAKE   *Geophis hoffmanni*
Pl. 156

SIZE: Largest male 241 mm TL (n = 2); largest female 233 mm TL (n = 1).

DISTRIBUTION: Caribbean slopes from Honduras to Panama; Pacific slopes from central Costa Rica to Panama.

IDENTIFYING FEATURES: Hoffmann's Earthsnake is uniform drab dark gray from head to tail. The head is dark brown, and the venter is light gray with dark gray spots anteriorly that grade to wide bands posteriorly. The light ventral coloring changes abruptly at the level of the ventral-most dorsal scales to the dark dorsal coloration. The most similar other species in shape and color are the Ridge-nosed Snake *(Amastridium veliferum)* and the White-headed Snake *(Enulius sclateri);* but the former has a gradual change in color from the venter to the dorsum, and the latter has a head that is distinctly lighter in color than the rest of the body.

SEXUAL DIMORPHISM: We know of no external features that distinguish the two sexes.

HABITAT: Like most members of the genus, this species is relatively bulky and is semifossorial, burrowing in leaf litter or loose soil.

DIET: Members of the genus *Geophis* eat earthworms, beetle larvae, and other invertebrates (Savage 2002).

REPRODUCTION: This species is presumed to lay eggs, like other members of the genus (Bogert and Porter 1966; Campbell et al. 1983).

TROPICAL SEEP SNAKE   *Hydromorphus concolor*
Pl. 157

SIZE: We have no measurements of this
snake from La Selva. As described by Savage
and Donnelly (1988), the male of this
species is as large as 690 mm TL and the female
as large as 797 mm TL.

DISTRIBUTION: Caribbean slopes from Guatemala to
Panama; Pacific slopes from central Costa Rica to Panama.

IDENTIFYING FEATURES: The dorsal ground color of this species is
tan (J.A. Campbell 1993, personal communication). The ventral-
most dorsal scales often have dark borders, and sometimes this
snake has scattered, small (one to two scales), dark spots on the
dorsum (Savage and Donnelly 1988). The head is uniform except
for a faint, postorbital, dark stripe. The chin scales are marked
with dark brown pigment, but the rest of the venter is light
(cream, yellow, or olive gray), often with scattered, small, dark
spots; the dark spots sometimes align to form a midventral, dark
stripe. The ventral coloration changes abruptly to the dorsal coloration at the ventral-most dorsal scale row (Savage and Donnelly 1988).

SEXUAL DIMORPHISM: We know of no external features that distinguish the two sexes.

HABITAT: This snake is rarely encountered at La Selva, probably
because of its aquatic habits. It has been described to inhabit leaf
litter accumulations in streams and along stream margins (Savage and Donnelly 1988).

DIET: This snake is presumed to eat small, soft-bodied prey (Savage and Donnelly 1988); Scott (1969) lists fish and frogs as prob-
able prey.

REPRODUCTION: We presume this species lays eggs.

REMARKS: A single museum specimen (CRE 7270) and an individual trapped by Catherine Pringle during aquatic sampling
(E.D. Brodie III 1995, personal communication) are the only
known records of this species from La Selva.

BROWN BLUNT-HEADED VINESNAKE *Imantodes cenchoa*
Pl. 158

OTHER COMMON NAMES: Bejuquillo,
Chunk-headed Snake, Blunt-headed
Tree Snake, Blunthead Tree Snake
SIZE: Largest male 1,200 mm TL (n = 39);
largest female 1,075 mm TL (n = 23).
DISTRIBUTION: Caribbean slopes from Tamaulipas,
Mexico, to Argentina and Paraguay; Pacific slopes from Chiapas, Mexico, to Ecuador.

IDENTIFYING FEATURES: This is one of the most commonly observed snakes at La Selva. It is medium in length, with an extremely thin body and a short, blunt head. The dorsal ground color is light tan, with bold, dark brown, diamond-shaped, mid-dorsal markings, each bordered by black. Typically, there is a U-shaped, dark brown mark on the head extending from the neck to the front of the interorbital region; this mark is sometimes reduced to a pair of oblong, parallel, dark blotches covering the parietal scales. Dark spots are found at the anterior tips of the U-shaped mark. The venter is light grayish brown. The pupil is vertical, and the iris is metallic tan. The bold dorsal markings distinguish this species from its congener, the Yellow Blunt-headed Vinesnake *(I. inornatus)*. No other snake at La Selva is as thin and has a blunt head and bold, diamond-shaped markings.

SEXUAL DIMORPHISM: We know of no external features that distinguish the two sexes. At La Selva, the two sexes are similar in mean total length (1,010 mm for the male; 960 mm for the female).

HABITAT: The species is arboreal and is found in primary and secondary forest. Active principally at night, this snake can be found foraging in shrubs and small trees along trails; by day it can be found coiled in bromeliads (Stuart 1948; Taylor 1951), under dead leaves, or between adpressed live leaves.

DIET: This rear-fanged snake consumes frogs and arboreal lizards that are hunted by careful visual and chemical (tongue) surveying of the thin branches of trees and shrubs. Anoles have been described as the primary diet item (Henderson and Nickerson 1976), but reptile eggs are also consumed (Landy et al. 1966), and frogs are consumed in captivity (Test et al. 1966).

REPRODUCTION: This species lays one to three eggs (usually two; Zug et al. 1979), and gravid females have been observed in most months of the year (Fitch 1970). At La Selva we observed females that appeared to be gravid on August 29, 1982, June 14, 1983, and July 25, 1983. Palpation revealed three putative oviductal eggs in two of these individuals and none in the third. A single specimen collected from the Braulio Carrillo National Park extension (April 8, 1986) had four unyolked follicles.

REMARKS: The species in the genus *Imantodes* are among the most docile of La Selva's snakes. They are not known to bite humans and can be positioned into a tight coil, where they remain motionless. The row of enlarged scales along the middle of the back is thought to provide support for the snake while it moves between distant supports in its arboreal habitat (Gans 1974).

YELLOW BLUNT-HEADED VINESNAKE

Imantodes inornatus

Pls. 159, 160

OTHER COMMON NAMES: Bejuquillo, Plain Tree Snake

SIZE: Largest male 978 mm TL (n = 6); largest female 950 mm TL (n = 10).

DISTRIBUTION: Caribbean slopes from Honduras to Panama; Pacific slopes from the Osa Peninsula, Costa Rica, to Ecuador.

IDENTIFYING FEATURES: This distinctively thin, midsized species has a yellowish tan dorsal ground color, shading to peach brown ventrally. Small, black flecks form very thin bands across the dorsum, but these are often irregular in shape and orientation. The head coloration is similar to that of the dorsum. The venter is yellow anteriorly shading to light peach posteriorly. Irregular, black flecks mark the belly, as does a thin, black midventral stripe. Like all members of the genus, the Yellow Blunt-headed Vinesnake is extremely thin and has a blunt head with large, protruding eyes. The pupils of the eyes are vertically elliptic. No other species of snake is as thin and has a blunt head and faded dorsal markings.

SEXUAL DIMORPHISM: We know of no external features that distinguish the two sexes. The adult male and female do not appear to differ in mean total length at La Selva (900 mm for the male; 910 mm for the female).

HABITAT: At La Selva, this species is nearly always found at night on vegetation near water (along streams or near swamps). It is often seen during the earliest storms of the rainy season.

DIET: This rear-fanged snake is known to eat frog eggs laid on leaves; it also eats adult frogs.

REPRODUCTION: This species lays eggs. A single female at La Selva laid three eggs, and a female from the Braulio Carrillo National Park extension contained seven unyolked follicles.

REMARKS: See remarks for the Brown Blunt-headed Vinesnake *(I. cenchoa)*.

TROPICAL MILKSNAKE   *Lampropeltis triangulum*
Pl. 161

OTHER COMMON NAMES: Coral Falsa, Tropical Kingsnake, Tropical King Snake, Milk Snake
SIZE: Largest male 1,820 mm TL (n = 6); largest female 1,485 mm TL (n = 2).
DISTRIBUTION: Atlantic and Caribbean slopes from Canada to Venezuela; Pacific slopes from Mexico to Ecuador.

IDENTIFYING FEATURES: This is the largest of the coralsnake mimics at La Selva. Additionally, the snake is not rear-fanged but is a constrictor and, therefore, the body is thicker than that of most other coralsnake mimics. The color pattern starts with a black head cap that extends posteriorly to the level of the temporal scales but may be interrupted by a narrow, white or yellow ring through the nasal scales of the snout. The first body ring is light (white or yellow) and covers the posterior parts of the temporal scales and the first dorsal scale rows. The second ring is black. The color pattern thereafter consists of a repetitive sequence of rings of the following colors: red (about six to 15 scale rows), black (about three to five scale rows), light (white or yellow; about two to three scale rows), and black. This pattern conforms to the tricolored dyad of Savage and Slowinski (1992). The tail is bicolored: black and light (white or yellow). In total, there are about nine red rings, 43 black rings, and 23 light rings. The scales of the red and light rings have black tips that cause the colors to fade in large individuals. Six other species of snakes at La Selva have a similar color pattern. Five of these, Allen's Coralsnake *(Micrurus alleni)*, the Central American Coralsnake *(M. nigrocinctus)*, the Central American Coralsnake Mimic *(Erythrolamprus mimus)*, the Skinkeater *(Scaphiodontophis annulatus)*, and the Calico Snake *(Oxyrhopus petolarius)*, differ by having red rings or bands bordered by light (white or yellow). The other similar species, the Tricolored Crowned Snake *(Tantilla supracincta)*, differs in being much smaller as an adult, having narrower black bands (one scale row), and possessing fewer red bands (less than 24).

SEXUAL DIMORPHISM: The adult male is longer than the female (Lee 1996), but no other external features aid in distinguishing the sexes.

HABITAT: The species is infrequently encountered but seems to prefer primary forest, where it is largely terrestrial. Although described as mostly nocturnal (Lee 1996), the three specimens that we recorded from La Selva were all found active during the day.

DIET: The adult eats a variety of vertebrates, including frogs, lizards, snakes, and mammals (Lee 1996); Scott (1969) also lists eggs (presumably of birds). At La Selva, rodents of the genus *Heteromys* were palped from two individuals. The species is famous for being able to consume large pitvipers.

REPRODUCTION: This species lays from five to 16 eggs (Fitch 1970).

REMARKS: When first handled this snake may jerk widely from side to side, cock the head to one side, and coil the tail, making it difficult to distinguish head from tail. Additionally, it may bury its head under coils when harassed.

SOUTHERN CAT-EYED SNAKE
Leptodeira annulata
Pl. 162

OTHER COMMON NAMES: Banded Cat-eyed Snake

SIZE: Largest male 540 mm TL (n = 2); largest female 729 mm TL (n = 2).

DISTRIBUTION: Caribbean slopes from Tamaulipas, Mexico, to Argentina; Pacific slopes from Guerrero, Mexico, to Ecuador.

IDENTIFYING FEATURES: This is the rarer of the two cat-eyed snakes at La Selva. It is medium in size. The dorsal ground color is brown, with a series of darker brown blotches down the middorsum; consecutive blotches are offset and at several places are fused to form a zigzag pattern. A series of smaller, dark brown blotches occurs along the side of the body. The first middorsal, dark blotch often is irregular in shape and projects forward to contact the parietal shields. The venter is immaculate light gray anteriorly, shading rapidly to a light peach posteriorly. The ventral surface of the tail has small, dark flecks. The head is brown with a dark, postorbital bar extending along the side of the face to the angle of the jaw. As with many nocturnal snakes, the pupil of the eye is vertical. This species is most similar to the Northern Cat-eyed Snake *(L. septentrionalis)*, from which it can be distinguished by the sections of zigzag markings on the dorsum.

SEXUAL DIMORPHISM: We know of no external features that distinguish the two sexes.

HABITAT: Too few specimens of this snake have been observed at La Selva to document habitat preferences. Scott (1969) considered it to be semiaquatic.

DIET: The first specimen of this species from La Selva was collected at the River Station, where it was attempting to regurgitate a small Marine Toad *(Bufo marinus)*. A small Marine Toad was palped from a second individual as well. Elsewhere this species is known to consume Vaillant's Frog *(Rana vaillanti)* (Ramirez et al. 1997) and other frogs and their eggs.

REPRODUCTION: This species is presumed to lay eggs.

NORTHERN CAT-EYED SNAKE   *Leptodeira septentrionalis*

Pls. 163, 164

OTHER COMMON NAMES: Cat-eyed Snake

SIZE: Largest male 785 mm TL (n = 32); largest female 958 mm TL (n = 25).

DISTRIBUTION: Caribbean slopes from southern Texas to Colombia; Pacific slopes from Sinaloa, Mexico, to Peru.

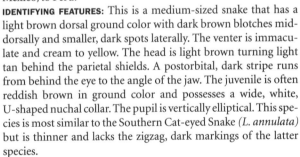

IDENTIFYING FEATURES: This is a medium-sized snake that has a light brown dorsal ground color with dark brown blotches mid-dorsally and smaller, dark spots laterally. The venter is immaculate and cream to yellow. The head is light brown turning light tan behind the parietal shields. A postorbital, dark stripe runs from behind the eye to the angle of the jaw. The juvenile is often reddish brown in ground color and possesses a wide, white, U-shaped nuchal collar. The pupil is vertically elliptical. This species is most similar to the Southern Cat-eyed Snake *(L. annulata)* but is thinner and lacks the zigzag, dark markings of the latter species.

SEXUAL DIMORPHISM: We know of no external features that distinguish the two sexes. No sex difference in the mean total length of the adult was noted for La Selva specimens (730 mm for the male; 770 mm for the female).

HABITAT: This commonly observed snake often is discovered at night on vegetation in secondary and primary forest. We have seen this snake frequently at night at the Research swamp and along the Research trail.

DIET: The species is rear-fanged and eats frogs and lizards. We palped the Marine Toad *(Bufo marinus),* the Olive Snouted Treefrog *(Scinax elaeochroa),* Fitzinger's Rainfrog *(Eleutherodactylus fitzingeri),* and Carpenter's Anole *(Norops carpenteri)* from snakes at La Selva. Elsewhere, this species is known to eat eggs of the frog genus *Agalychnis* (photo in Greene 1997).

REPRODUCTION: This species lays from seven to 12 eggs (Ditmars 1939; Haines 1940). Elsewhere, gravid females have been observed in every month of the year (Oliver 1947). A single specimen from the Braulio Carrillo National Park extension had 17 unyolked follicles.

REMARKS: Two species that look generally similar to the Northern Cat-eyed Snake—the Diamondback Racer *(Drymobius rhombifer)* and Degenhardt's Scorpion-eater *(Stenorrhina degenhardtii)* —are known from the lowland wet forests of Costa Rica. Therefore, individuals that appear similar but do not fit the detailed description of the Northern Cat-eyed Snake should be compared with the descriptions for the Diamondback Racer and Degenhardt's Scorpion-eater (see under "Additional Species").

GREEN PARROTSNAKE   *Leptophis ahaetulla*
Pls. 165, 166

OTHER COMMON NAMES: Lora, Lora Falsa, Green Parrot-snake, Parrotsnake, Northern Green Frogger

SIZE: Largest male 1,980 mm TL (n = 14); largest female 1,595 mm TL (n = 12).

DISTRIBUTION: Caribbean slopes from Veracruz, Mexico, to Argentina; Pacific slopes from central Costa Rica to Ecuador.

IDENTIFYING FEATURES: This species is the largest of the parrotsnakes at La Selva. It has a dorsum that is uniform dull lime green, broken only by a series of 12 thin, black stripes formed by black pigment along the keeled dorsal scale rows. The head is light green with a thin, black line from the tip of the snout to the eye followed by a wide, black stripe (faded in large individuals) from the eye to the posterior end of the jaws. This species is similar in appearance to the Short-nosed Vinesnake *(Oxybelis brevi-*

rostris) and the Green Vinesnake *(Oxybelis fulgidus)* but differs from these in having large, laterally oriented eyes rather than forward-oriented eyes and a grooved snout. The Green Parrotsnake can be distinguished most reliably from the Satiny Parrotsnake *(Leptophis depressirostris)* by the absence of a loreal scale. Additionally, the Green Parrotsnake is a lighter, duller green than the Satiny Parrotsnake.

SEXUAL DIMORPHISM: We know of no external features that distinguish the two sexes. At La Selva, the adult male averages a slightly longer total length (1,780 mm) than the female (1,490 mm).

HABITAT: Although this species can be found in open areas, it tends to be more common in areas of second growth. Like all parrotsnakes, this is an active, diurnal, and arboreal animal.

DIET: The species is a frog specialist. At La Selva, this snake is known to consume the Common Mexican Treefrog *(Smilisca baudinii)* and nestling birds. We saw two snakes locate Common Mexican Treefrogs in trees, where the frogs presumably were roosting during the day. Ramirez et al. (1997) report Vaillant's Frog *(Rana vaillanti)* as a prey item.

REPRODUCTION: This snake lays eggs. Elsewhere gravid females are known from every month of the year (Iquitos, Peru; Oliver 1947). Rand (1969) reports fresh eggs and old shells of this species together in bromeliads, suggesting that this species reuses nest sites.

REMARKS: Individuals are aggressive upon capture and bite readily.

SATINY PARROTSNAKE   *Leptophis depressirostris*
Pls. 167, 168
OTHER COMMON NAMES: Lora, Parrotsnake
SIZE: Largest male 1,595 mm TL (n = 32); largest female 1,220 mm TL (n = 32).
DISTRIBUTION: Caribbean slopes from Nicaragua to Panama; Pacific slopes from southwestern Costa Rica to Ecuador.

IDENTIFYING FEATURES: This species is the most common of the parrotsnakes. The adult is uniform bright green; the juvenile is bronzy green shading to bright lime green laterally. In large adults, a pair of thin, black stripes occurs along the keels of the paravertebral scales. A thin, black stripe extends from the tip of the snout to the eye; this is followed by a broad, black stripe from the eye to the posterior end of the jaw. The venter is light yellowish green, a feature that distinguishes the species from the Striped Parrotsnake *(L. nebulosus)*. The Satiny Parrotsnake differs from the Green Parrotsnake *(L. ahaetulla)* in having a loreal scale between the eye and external nares, fewer black, paravertebral stripes, and darker, brighter green dorsal coloration.

SEXUAL DIMORPHISM: No external features distinguish the sexes, and mean adult size does not differ between the sexes for animals from La Selva (1,060 mm TL for the male; 1,000 mm for the female).

HABITAT: Virtually all of the specimens that we have seen have come from disturbed areas, usually the laboratory clearing. This snake is active and diurnal, and climbs readily.

DIET: This species is rear-fanged and eats frogs. Diet items known from La Selva include Bransford's Litterfrog *(Eleutherodactylus bransfordii)* and the juvenile Smoky Jungle Frog *(Leptodactylus pentadactylus)*.

REPRODUCTION: This species lays eggs. For such a common snake, virtually nothing is known of the timing of reproduction. We observed a pair mating on January 16, 1983. Of two females from the Braulio Carrillo National Park extension, one contained four oviductal eggs and the other four yolking follicles (5 and 12 April, 1986, respectively). Dundee and Liner (1974) described a clutch of three eggs, along with old shells, located in a bromeliad; the eggs hatched during July 1973. The observation of eggs and old shells in the same nest suggests that the snake reuses nest sites.

REMARKS: This snake is quite aggressive and delivers a slashing bite by rotating the maxilla so that the rear fangs are exposed. An anticoagulant in the saliva causes profuse bleeding in humans who are bitten, but we have experienced no other adverse effects.

STRIPED PARROTSNAKE
Leptophis nebulosus

Pl. 169

OTHER COMMON NAMES: Lora, Bronze-striped Parrotsnake

SIZE: Largest male 898 mm TL (n = 4); largest female 1,872 mm TL (n = 14).

DISTRIBUTION: Caribbean slopes from northeastern Honduras to Costa Rica; Pacific slopes of southwestern Costa Rica.

IDENTIFYING FEATURES: This snake is the rarest of the three species of parrotsnakes at La Selva, although it is not uncommon. The middorsum has a wide, bronzy stripe (greater than three scale rows), whereas the lateral areas are either lime green or bluish green. The supralabials (upper lip scales) are yellowish above and whitish below. The head is uniform green with a thin, black stripe from the tip of the snout through the eye to the edge of the posterior-most supralabial. The venter is bronzy grayish brown, a feature that distinguishes the species from the Satiny Parrotsnake *(L. depressirostris)*. Like all parrotsnakes, the Striped Parrotsnake can be differentiated from species of the vinesnake genus *Oxybelis* by the laterally oriented eyes and ungrooved snout.

SEXUAL DIMORPHISM: No external features distinguish the two sexes. At La Selva, the mean total length of the adult male is shorter (850 mm) than that of the female (1,050 mm).

HABITAT: This is an active, diurnal snake that can be found in open secondary forest and deep primary forest.

DIET: Frogs are the major diet item of this rear-fanged species. At La Selva, this snake consumes the Olive Snouted Treefrog *(Scinax elaeochroa)*.

REPRODUCTION: This species is presumed to lay eggs, but nothing is known of the timing of reproduction.

FIRE-BELLIED SNAKE   *Liophis epinephalus*

Pls. 170, 171

SIZE: Largest male 413 mm TL (n = 1); largest female 595 mm TL (n = 1). Elsewhere in Costa Rica this species reaches 770 mm TL.

DISTRIBUTION: Caribbean slopes from Costa Rica to Panama; Pacific slopes from Costa Rica to Peru.

IDENTIFYING FEATURES: This snake is medium in size and is banded with dark gray or black (two to three scale rows) and reddish brown (five to six scale rows). The head is uniform dark brown or black except for the ventral half of the supralabials, which are immaculate white. The ventral-most dorsal scales are brown or dark gray, separating the dorsal from the ventral coloration. The venter is tomato red with rectangular, black markings laterally. When disturbed, the Fire-bellied Snake flattens its body and displays light blue and reddish skin between the dorsal scales, which creates a crude coral-mimicking pattern. The Fire-bellied Snake is not likely to be confused with any other species of snake at La Selva.

SEXUAL DIMORPHISM: We know of no external features that distinguish the two sexes.

HABITAT: Too few specimens of this snake have been observed at La Selva to allow characterization of its habitat. This species may be more common at higher elevations. Scott (1969) considered it to be semiaquatic.

DIET: This snake is rear-fanged and eats frogs. The species is unusual in being able to consume dart-poison frogs of the genus *Phyllobates* without succumbing to the frogs' toxins (Myers and Daly 1983).

REPRODUCTION: This species lays eight to nine eggs (Leenders 2001). An adult female from the Braulio Carrillo National Park extension had seven yolking follicles.

REMARKS: Dixon (1980) assigns this species to the genus *Liophis;* older literature places this snake in the genus *Leimadophis.*

SALMON-BELLIED RACER   *Mastigodryas melanolomus*
Pls. 172, 173

OTHER COMMON NAMES: Orange-bellied Racer
SIZE: Largest male 1,099 mm TL (n = 12); largest female 1,402 mm TL (n = 19).
DISTRIBUTION: Caribbean slopes from Tamaulipas, Mexico, to Panama; Pacific slopes from Nayarit, Mexico, to Panama.

IDENTIFYING FEATURES: The color pattern of this midsized species is dimorphic between the adult and juvenile. The adult has a dorsum that is uniform light gray brown. A pair of faint, dorsolateral, light stripes (one to two scale rows wide) can be seen upon close inspection; these are bordered below by darker gray brown. On the venter, this snake is immaculate salmon red, and on the head, it is gray brown with a dark eye stripe from the tip of the snout, through the eye, to the angle of the jaw. The supralabials are cream with irregular, gray blotches. The juvenile differs from the adult in possessing a series of dark brown blotches separated by light blue gray to white bands, creating a nearly banded appearance. The bands sometimes fade from posterior to anterior, so that only the posterior half of the body appears to be banded. The venter is pinkish to salmon and not as bright as in the adult. The head is uniform gray brown with a dark eye stripe. The juvenile can be confused with the Brown Forest Racer *(Dendrophidion percarinatum)* and the Lowland Forest Racer *(D. vinitor)* but differs from these species in the color of the venter, which is white or yellow in these other two species.

SEXUAL DIMORPHISM: No external features distinguish the two sexes. At La Selva, the male and female are approximately the same mean total length (970 mm for the male; 1,090 mm for the female).

HABITAT: This is the common diurnal racer of primary and secondary forests. It can sometimes be found coiled low in vegetation at night.

DIET: The diet of this snake consists of lizards and frogs. Diet items known from La Selva include the Litter Skink *(Sphenomorphus cherrei)* and the Central American Whiptail *(Ameiva festiva).* Scott (1969) lists frogs, lizards, and probably insects as prey items, and Ramirez et al. (1997) list Vaillant's Frog *(Rana vaillanti).*

REPRODUCTION: Presumably this snake lays eggs.

REMARKS: This animal readily bites upon capture. Some authors (e.g., Lee 1996) place this species in the genus *Dryadophis.*

SPOTTED WOODSNAKE *Ninia maculata*
Pls. 174, 175

OTHER COMMON NAMES: Spotted Wood Snake
SIZE: Largest male 215 mm TL (n = 1); largest female 260 mm TL (n = 1). Leenders (2001) reports individuals up to 352 mm TL.
DISTRIBUTION: Caribbean slopes from Honduras to Panama.

IDENTIFYING FEATURES: This small snake has a dorsum that is dark bluish gray in ground color, with a series of black, irregularly shaped bands that are two to three scale rows wide. These bands are often disconnected near the middorsum, occasionally creating a checkered pattern. The venter has a checkerboard pattern of light gray and black, square or rectangular markings; these markings occasionally align on the midventer to create a dark stripe. The head is gray with a black nuchal collar, bordered anteriorly by light gray. The latter mark is disrupted middorsally by black. The labials are white with dark gray sutures, giving the lips a barred appearance. The light and dark markings of the juvenile are bold and become faded in the adult. This snake has a rough appearance because the scales are strongly keeled. Additionally, a middorsal ridge extends along the entire length of the body, creating a body cross section similar to a triangular file. The only similarly colored species at La Selva is the Cloudy Slugeater *(Sibon nebulatus)*, which differs from the Spotted Woodsnake in having smooth scales and a rounded body in cross section.

SEXUAL DIMORPHISM: We know of no external features that distinguish the two sexes.

HABITAT: Because of its secretive habits, this leaf litter snake is unlikely to be encountered. It is found in primary and secondary forest where litter accumulates. It was the most abundant snake captured in pitfall traps during the leaf litter study at La Selva summarized by Lieberman (1986).

DIET: Members of the genus are presumed to eat slugs, worms, and other invertebrates (Scott 1969).

REPRODUCTION: Presumably this species is oviparous.

REMARKS: When disturbed this snake flares the quadrate and squamosal bones to widen the back of the skull, mimicking the head shape of pitvipers. This may be followed by a series of death-feigning behaviors, such as tongue extrusion and loss of the righting response when overturned.

RED COFFEESNAKE   *Ninia sebae*
Pls. 176, 177

OTHER COMMON NAMES: Redback Coffeesnake, Redback Coffee Snake
SIZE: Largest male 330 mm TL (n = 1); largest female 325 mm TL (n = 1).
DISTRIBUTION: Caribbean slopes from Veracruz, Mexico, to Costa Rica; Pacific slopes from Oaxaca, Mexico, to Costa Rica.

IDENTIFYING FEATURES: This small snake has a brick red dorsal ground color, with the exception of the head and neck region. Here, bold black and light yellow or white bands are found. The head cap is black and is bordered posteriorly by a light nuchal collar (two to three scale rows wide), followed by a second wide, black band (five to seven scale rows wide) that is bordered posteriorly by a narrow, light band (two to three scale rows wide). The venter is immaculate white. This species is similar in color pattern to the juvenile Clelia *(Clelia clelia)*. However, the Red Coffeesnake differs from the juvenile Clelia in having two light bands at the anterior end (rather than one), heavily keeled dorsal scales (rather than smooth), and a distinctive middorsal, bony ridge that gives the snake the shape of a triangular file (rather than being round in cross section).

SEXUAL DIMORPHISM: During the breeding season, the male develops tubercles on the chin (Lee 1996). No size dimorphism has been described for the species.

HABITAT: This snake is secretive and lives in the leaf litter of pastures, agricultural areas, second growth, and primary forest.

DIET: This species has been described to eat earthworms, slugs, snails, and leeches (Greene 1975; Seib 1985b). To this list we add a Northern Masked Rainfrog *(Eleutherodactylus mimus)* palped from a snake at La Selva.

REPRODUCTION: This snake is oviparous, laying from one to four eggs during the wet season (Lee 1996).

REMARKS: This species was not known from La Selva until 1983, when one was palped from the stomach of a Central American Coralsnake Mimic *(Erythrolamprus mimus).*

RUGOSE LITTERSNAKE   *Nothopsis rugosus*
Pl. 178

SIZE: Largest male 380 mm TL (n = 5); largest female 420 mm TL (n = 6).

DISTRIBUTION: Caribbean slopes from Honduras to Panama; Pacific slopes from the Osa Peninsula, Costa Rica, to Colombia.

IDENTIFYING FEATURES: This odd, small snake differs from all other La Selva colubrids in that the top of the head is covered with small scales rather than large, symmetrical plates. The dorsum is gray with a variegated pattern of dark gray and brown, often tinged laterally with yellow. A series of dark gray, triangular markings extends along the side. These markings are bordered by white or light yellow and often connect along the sides to form a series of X-shaped structures. The tail is tawny, and the head is uniform dark gray. The venter is gray with paired brown spots on each scute except for the anterior-most ones. Although the color pattern is vaguely reminiscent of the Terciopelo *(Bothrops asper),* the Rugose Littersnake is not likely to be confused with that species or any other.

SEXUAL DIMORPHISM: We know of no external features that distinguish the two sexes. We observed little difference in the mean total lengths of the adult male (350 mm) and female (380 mm) at La Selva.

HABITAT: This species at La Selva has been observed day and night but is secretive, living under leaf litter in forested areas.

DIET: At La Selva, this snake consumes Bransford's Litterfrog (*Eleutherodactylus bransfordii*), and we presume that it is a member of the assemblage of small, leaf litter snakes that eat mostly frogs. Scott (1969) records a salamander as a diet item in one individual.

REPRODUCTION: This species lays eggs. We palpated five females with two to four putative oviductal eggs (April 27, 1982; September 18, 1982; July 5, 1983; March 7, 1984; June 26, 1984). Because these dates span both wet and dry season months, we infer that reproduction occurs year-round.

REMARKS: This snake is extremely sluggish when held in the hand and is unlikely to be very active while ranging freely. It allows itself to be contorted into bizarre loops and remains so for long periods of time.

BROWN VINESNAKE   *Oxybelis aeneus*
Pl. 179

OTHER COMMON NAMES: Bejuquillo, Mexican Vine Snake, Brown Vine Snake
SIZE: Largest male 1,420 mm TL (n = 8); largest female 1,495 mm TL (n = 10).
DISTRIBUTION: Caribbean slopes from Tamaulipas, Mexico, to Bolivia; Pacific slopes from southern Arizona to Peru.
IDENTIFYING FEATURES: This snake, like all members of the genus, has binocular vision made possible by the forward-directed eyes and a slight depression from the tip of the snout to the orbit. The iris is light yellow dorsally and ventrally and dark brown anteriorly and posteriorly. The body is coppery brown, interrupted by a lattice of black flecks or brown spots. The venter is white shading to gray and gray brown posteriorly. The head is coppery brown dorsally, changing abruptly to off-white at the supralabials (upper lip scales). The head is extremely thin and elongate, and the inside of the mouth is deep purple to black. This is La Selva's

only diurnal, brown vinesnake, which distinguishes it from the more blunt-headed, nocturnal, brown vinesnakes of the genera *Leptodeira* and *Imantodes,* as does the unique coloration of the mouth cavity.

SEXUAL DIMORPHISM: No external features distinguish the two sexes. The adult male from La Selva averages a slightly shorter total length (1,284 mm) than the female (1,390 mm).

HABITAT: All individuals that we have seen from La Selva have come from open second-growth areas or forest edge along major rivers. This animal is active by day.

DIET: This species is a rear-fanged lizard specialist, most frequently eating anoles (Henderson 1982) but also consuming frogs, birds, and insects (Henderson and Binder 1980). At La Selva, all three diet items that have been observed were Central American Whiptails *(Ameiva festiva).*

REPRODUCTION: This snake lays eggs. We palpated four eggs in a female on June 16, 1983. Reproduction in this species may be restricted to the dry season (Censky and McCoy 1988). Eggs are 50 mm in length and take 2.5 months to hatch (Leenders 2001).

REMARKS: When hunting, this snake may bob its head and body forward and backward in an oscillating fashion while slowly moving forward. This movement mimics the appearance of wind-blown vines, potentially hiding the snake from its prey (Fleishman 1985). This snake gapes and lunges when threatened, and bites to humans can result in a blistered swelling (Crimmins 1937). Ritualized male-male combat has been observed in this snake (Leenders 2001).

SHORT-NOSED VINESNAKE   *Oxybelis brevirostris*
Pl. 180

OTHER COMMON NAMES: Bejuquillo, Short-nosed Vine Snake

SIZE: Largest male 950 mm TL (n = 1); largest female 1,025 mm TL (n = 2).

DISTRIBUTION: Caribbean slopes from Honduras to Peru; Pacific slopes from Panama to Ecuador.

IDENTIFYING FEATURES: This vinesnake is medium in size and uniform olive green dorsally shading to brighter green laterally. The head is olive green with a black eye mask extending from the tip of the snout, through the eye, to the level of the posterior tempo-rals. The venter is lime green. The eyes face forward and have binocular vision, aided by a slight depression along the snout from the nares to the orbit. The species is most likely to be con-fused with the Green Vinesnake *(O. fulgidus)*—from which it can be distinguished by the lack of a bold, light, ventrolateral stripe—and the parrotsnakes of the genus *Leptophis,* from which it can be distinguished by the forward-directed eyes and grooved snout.

SEXUAL DIMORPHISM: We know of no external features that distin-guish the two sexes.

HABITAT: This diurnal, arboreal snake probably prefers open sec-ond growth and forest edge. It may be more common at higher elevations in the Braulio Carrillo National Park.

DIET: This snake is an active forager and, like its congeners, is as-sumed to eat principally lizards.

REPRODUCTION: This snake lays eggs. One female from the Braulio Carrillo National Park extension had three oviductal eggs.

REMARKS: This snake may present an aggressive display when dis-turbed, during which it raises its head in an S-shaped coil, opens its mouth, and flattens itself in the neck region (Leenders 2001).

GREEN VINESNAKE
Oxybelis fulgidus

Pl. 181

OTHER COMMON NAMES: Bejuquillo, Lora

SIZE: Largest female 1,875 mm TL (n = 1).

DISTRIBUTION: Caribbean slopes from Veracruz, Mexico, to Argentina; Pacific slopes from the Isth-mus of Tehuantepec, Mexico, to Panama.

IDENTIFYING FEATURES: This green vinesnake has been ob-served only a few times at La Selva. The dorsal ground color is bright leaf green. The head is elongate and narrow, with a shallow groove extending from the orbits to the tip of the snout, allowing the large, anteriorly oriented eyes to attain binocular vision. A thin, black stripe extends from the tip of the snout through the orbit. Posteriorly, the stripe is broad and passes through the angle of the jaw. The uniform dorsal coloration is interrupted on the side of the body, where a distinct ventrolateral, white stripe is pre-sent, extending from the supralabials through most of the length of the body. This species can be distinguished from the Short-nosed Vinesnake *(O. brevirostris)* and all species of the genus *Lep-tophis* by the presence of the ventrolateral stripe.

SEXUAL DIMORPHISM: No external features distinguish the two sexes.

HABITAT: The species is arboreal and diurnal. Too few specimens have been captured from La Selva to determine its local habitat preferences. It has been described as preferring open second growth and forest edge (Lee 1996).

DIET: This snake is rear-fanged and consumes vertebrates, mostly lizards (Henderson 1982). The single specimen that we observed at La Selva had consumed a Central American Whiptail *(Ameiva festiva)*.

REPRODUCTION: This species lays up to 10 eggs per clutch (Conners 1989).

CALICO SNAKE   *Oxyrhopus petolarius*

Pls. 182, 183

OTHER COMMON NAMES: Coral Falsa, Calico False Coralsnake

SIZE: Largest male 782 mm TL (n = 3); largest female 1,035 mm TL (n = 11).

DISTRIBUTION: Caribbean slopes from Veracruz, Mexico, to northern Brazil; Pacific slopes from central Costa Rica to Ecuador.

IDENTIFYING FEATURES: This species is a medium-sized coralsnake mimic possessing a pattern of light and dark bands that do not encircle the body. The head cap is black, bordered posteriorly by a U-shaped, light (usually white, occasionally orange) nuchal collar that covers the posterior two-thirds of the parietal scales. The anterior part of the body has alternating black and light bands. The light bands are three to four scale rows wide and may be white or yellow or light red bordered by white or yellow. The black bands are the widest of the bands and are wider anteriorly (about 11 scale rows) than posteriorly (about four scale rows). The colors throughout the length of the body appear faded. Often one to several of the light bands are incomplete, forming V- or Y-shaped, dark markings across the dorsum. Rarely, several of these irregularities join together to form a black zigzag pattern. The pupil is vertical. The color pattern mimics tricolored (occasionally) or bicolored (commonly) coralsnakes. This species is easily distinguished from true coralsnakes by the possession of bands rather than rings, the distinct color pattern in the neck region, and the presence of large, easily visible eyes. The only other bicolored mimic at La Selva is the Halloween Snake *(Urotheca euryzona)*, which has rings, with the light ones being two-toned (orange dorsally, white ventrally). The tricolored morph looks similar to the Central American Coralsnake Mimic *(Erythrolamprus mimus)*, the Tropical Milksnake *(Lampropeltis triangulum)*, and the Tricolored Crowned Snake *(Tantilla supracincta)*. The first differs by having fewer red rings (12 to 14), brighter colors, and red rings that are wider than the black rings; the other two species differ by having red rings or bands that contact the black ones.

SEXUAL DIMORPHISM: We know of no external features that distinguish the two sexes.

HABITAT: This species is active day and night in primary and secondary forests, although most of our daytime observations are from early morning.

DIET: This snake is rear-fanged and eats lizards and snakes (Lee 1996). One specimen at La Selva was observed to bite and constrict a Central American Whiptail *(Ameiva festiva)*. Scott (1969) records mammal hair in the fecal matter of this species.

REPRODUCTION: This species lays from five to 10 eggs (Fitch 1970), but little is known of the timing of reproduction.

REMARKS: We follow Savage (2002) for the spelling of the specific epithet; other recent authors refer to this snake as *O. petola*.

BIRD-EATING SNAKE *Pseustes poecilonotus*
Pls. 184, 185

OTHER COMMON NAMES: Sabanera, Mica, Mahogany Rat Snake, Puffing Snake, Bird Snake

SIZE: Largest male 2,010 mm TL (n = 18); largest female 2,020 mm TL (n = 18).

DISTRIBUTION: Caribbean slopes from San Luis Potosí, Mexico, to Brazil; Pacific slopes from Oaxaca, Mexico, to Ecuador.

IDENTIFYING FEATURES: This is a large snake with a ground color of olive green in the adult; this color is disrupted by irregular, indistinct, red, yellow, or gray brown bars (two to six scale rows wide), each edged by dark brown or black. The head coloration is dark grayish green, changing abruptly to white or yellow on the upper lip; the light color on the upper lip is confluent with the white or yellow chin. The venter is yellowish anteriorly and gray green posteriorly. Unlike the adult, the juvenile is banded dark olive green and brown, possesses a dark brown, postorbital bar (from the back of eye to the level of first dorsal scales), and has a dark brown blotch on the supralabials below each eye. The iris is tawny brown with darker brown blotches anterior and posterior to the pupil. This species is not likely to be confused with any other.

SEXUAL DIMORPHISM: No external features distinguish the two sexes. The mean total length of males captured at La Selva averaged slightly less (1,540 mm) than that of females (1,720 mm).

HABITAT: This snake is diurnal and often climbs trees and shrubs in primary and secondary forests.

DIET: This species has teeth on the maxilla that are slightly enlarged posteriorly and is known to eat birds and bird eggs (Alvarez del Toro 1983; Sexton and Heatwole 1965). At La Selva, four bats were palped from the stomach of one individual, and Leenders (2001) observed this snake eating a White-collared Swift (*Streptoprocne zonaris*).

REPRODUCTION: We presume this species lays eggs.

REMARKS: When confronted by humans, this snake flattens its neck laterally, inflates its lung, gapes its mouth, and strikes.

PINK-BELLIED LITTERSNAKE   *Rhadinaea decorata*
Pl. 186

OTHER COMMON NAMES: Vibora de Sangre, Adorned Graceful Brown Snake, Red-bellied Littersnake

SIZE: Largest male 385 mm TL (n = 23); largest female 443 mm TL (n = 19).

DISTRIBUTION: Caribbean slopes from San Luis Potosí, Mexico, to Panama; Pacific slopes from the Osa Peninsula, Costa Rica, to Ecuador.

IDENTIFYING FEATURES: This is one of the more commonly seen small snakes. The dorsum is brown with a pair of dorsolateral, light stripes, each bordered above and below by thin, dark stripes. The light stripes are bold and easy to see for the anterior-most fourth to third of the body length, after which they fade abruptly. Another light stripe extends along the ventral-most dorsal scale row on either side of the body. The venter is light gray anteriorly shading to tangerine posteriorly. A series of small, dark spots along the lateral-most edge of each ventral scute separates the light ventral coloration from the lateral, light stripe. The head is brown with a pair of postorbital, light spots, each bordered by black. The species is similar to the Long-tailed Littersnake *(Urotheca decipiens)* and the Orange-bellied Littersnake *(U. guentheri)* at La Selva but can be distinguished from them by the fading of the light stripes toward the posterior end of the body. The genera *Tantilla* and *Trimetopon* are also small, brown littersnakes but these differ from the Pink-bellied Littersnake in that they lack light stripes and possess tan or white nuchal collars or cheek patches.

SEXUAL DIMORPHISM: No external features distinguish the two sexes; the mean total lengths of the male and the female at La Selva are similar (330 mm and 360 mm, respectively).

HABITAT: The Pink-bellied Littersnake is found in primary and secondary forests and usually is observed for only a brief moment as it slithers off a trail and under the leaf litter. It is active by day.

DIET: This snake consumes frogs (Scott 1983d) and salamanders (Smith 1943). At La Selva it eats Bransford's Litterfrog *(Eleutherodactylus bransfordii)* and the Common Tink Frog *(E. diastema)*.

REPRODUCTION: This species lays eggs, but nothing is known of the timing of reproduction. We palpated two gravid females on August 12 and 21, 1982, one of which appeared to contain three eggs. A third female laid one egg while in captivity on July 2, 1983. A hatchling-sized animal (145 mm TL) was observed by us on March 5, 1983.

REMARKS: The Spanish common name, Vibora de Sangre, comes from a folk legend that suggests that people bitten by this snake bleed to death. This is a common tale associated with snakes possessing red venters.

SKINKEATER   *Scaphiodontophis annulatus*
Pls. 187, 188

OTHER COMMON NAMES: Neck-banded Snake, Shovel-toothed Snake, Guatemalan Neckband Snake

SIZE: Largest male 445 mm TL (n = 4); largest female 685 mm TL (n = 4).

DISTRIBUTION: Caribbean slopes from Tamaulipas, Mexico, to northern Colombia; Pacific slopes from the Isthmus of Tehuantepec, Mexico, to Panama.

IDENTIFYING FEATURES: This midsized species is unusual in that the color pattern changes dramatically from the juvenile to the adult. At hatching, this snake possesses a bone white head cap that extends posteriorly to include the anterior two-thirds of the parietal shields. The ground color of the rest of the body is dark gray to black. This coloring is disrupted only by thin, white bands across the top half of the dorsum. The adult is a mimic of tricolored coralsnakes, but with a dorsal pattern of bands that do not continue onto the belly. The head cap is gray anteriorly shading to white or yellow between the orbits. The black bands begin at the back of the head and neck. The body consists of a repetitive series of red, light (gray, white, or yellow), black, and light bands. This constitutes the tricolored monad pattern of Savage and Slowinski (1992). The red bands are seven to nine scale rows wide, the light bands two to three, and the black bands four to seven. The scales of the red bands are each tipped in black, and there are 10 total red bands along the length of the body. The bands are often offset on each side, creating irregularities at the midline. The iris is golden, and the large eye is easily visible in the head cap. The sequence of dorsal colors is identical to Allen's Coralsnake *(Micrurus alleni)* and the Central American Coralsnake *(M. nigrocinctus)*. However, these species differ from the Skinkeater in being ringed, having a small, black iris that is difficult to discern in the black skullcap, and possessing more than 15 red rings. Four other colubrids share the general color pattern of the Skinkeater. However, two of them, the Tropical Milksnake *(Lampropeltis triangulum)* and the Tricolored Crowned Snake *(Tantilla supracincta),* have red rings or bands bordered by black ones, and the other two, the Central American Coralsnake Mimic *(Erythrolamprus mimus)* and the Calico Snake *(Oxyrhopus petolarius),* have more than 10 red rings or bands.

SEXUAL DIMORPHISM: No external features distinguish the two sexes. Mean total length is shorter in the male (364 mm) than the female (547 mm) at La Selva.

HABITAT: This species lives in and under fallen leaves and can be found in primary and secondary forests with thick leaf litter. It is active during the day.

DIET: An unusual adaptation allows the teeth to bend where they attach to the jaw. This mechanism has been suggested to allow this snake to grasp lizards (Savitzky 1981). Others have described this genus as specializing on skinks, especially of the genus *Sphe-*

nomorphus (Henderson 1984); however, we have also observed the Ground Anole *(Norops humilis)* and the Common Tink Frog *(Eleutherodactylus diastema)* in the diet of this snake.

REPRODUCTION: The female lays three to four eggs in the rainy season (Alvarez del Toro and Smith 1958; Henderson 1984).

REMARKS: We follow Savage and Slowinski (1996) for the specific epithet; older literature refers to this snake as *S. venustissimus.* The tail of the species is unusually long and easily broken (Slowinski and Savage 1995) and may break more than once. The snake sometimes thrashes its tail when threatened, apparently to attract predators to this part of the body (Henderson 1984).

RINGED SLUGEATER *Sibon annulatus*
Pl. 189

OTHER COMMON NAMES: Ringed Snail-eater, Ringed Slug-eater

SIZE: Largest male 550 mm TL (n = 2); largest female 555 mm TL (n = 2).

DISTRIBUTION: Caribbean slopes from Honduras to Panama.

IDENTIFYING FEATURES: Like all members of the genus, this species is a medium-sized, blunt-headed snake. The color pattern consists of irregular, dark reddish brown rings that are separated dorsally by irregular tan rings that shade to darker greenish brown laterally and then change abruptly to cream on the ventral-most dorsal scales and the ventral scutes. The venter is light and dark; the rings are often offset midventrally. The dorsal rings usually become less regular posteriorly. The pupil is vertical. This species is most similar to the Bicolored Snaileater *(Dipsas bicolor),* and differs from that species in lacking orange and in having green on the light rings.

SEXUAL DIMORPHISM: We know of no external features that distinguish the two sexes.

HABITAT: The Ringed Slugeater can be found at night on vegetation in forested areas. The species appears to be rare at La Selva.

DIET: As with all members of the genus, this species eats slugs.

REPRODUCTION: This snake is presumed to lay eggs.

REMARKS: This species is a member of the "goo" snakes, an assemblage of snakes that all eat soft-bodied invertebrates that appear as "goo" when stomach contents are examined for diet studies.

Related to this, the scat of this species (and probably others in the genus) is an amorphous, dark blob. When harassed, this snake flares the quadrate and squamosal bones at the back of the skull to create a triangular head shape. However, the snakes of this genus are extremely docile and are not known to bite humans.

LICHEN-COLORED SLUGEATER
Sibon longifrenis
Pl. 190

OTHER COMMON NAMES: Lichen-colored Snail-eater, Mottled Slugeater

SIZE: Largest male 610 mm TL (n = 3); largest female 573 mm TL (n = 5).

DISTRIBUTION: Caribbean slopes from Honduras to Panama.

IDENTIFYING FEATURES: This midsized, blunt-headed snake has a color pattern similar to the green morph of the Eyelash Viper *(Bothriechis schlegeli)* but lacks the large, triangular head and projecting superciliary (eyelash) scales of that species. The Lichen-colored Slugeater has an olive green ground color with light brown, middorsal blotches that are outlined by dark brown. The blotches are typically wider than they are long anteriorly, whereas the length and width of the blotches are equal posteriorly. The head is mottled with green, brown, and black. Ventrolaterally, there is a series of white blotches separated from one another by dark-bordered, light brown blotches. The venter is yellow with fine, black punctuations. The iris is olive, and the pupil is vertical.

SEXUAL DIMORPHISM: No external features distinguish the two sexes. The mean total lengths of the adult male and female at La Selva are similar (590 mm and 500 mm, respectively).

HABITAT: This nocturnal, arboreal species is observed most often on vegetation in and surrounding the Research and Cantarana swamps. It has not been observed far from water.

DIET: This species eats slugs.

REPRODUCTION: This snake is presumed to lay eggs.

REMARKS: See remarks for the Ringed Slugeater *(S. annulatus).*

CLOUDY SLUGEATER
Sibon nebulatus
Pl. 191

OTHER COMMON NAMES: Dormilona,
Cloudy Snail Sucker
SIZE: Largest male 795 mm TL (n = 12);
largest female 799 mm TL (n = 15).
DISTRIBUTION: Caribbean slopes from Veracruz,
Mexico, to Brazil; Pacific slopes from Nayarit, Mexico,
to Ecuador.

IDENTIFYING FEATURES: This is a medium-sized snake with a short, thick head. The dorsal ground color is purplish gray, with an irregular banding of dark gray or black; the bands are often bordered by white spots and flecks less than one scale row wide. The venter is white suffused with dark gray to brown flecks, and with broken, dark gray bands shading to black posteriorly. The head is dark brown to black suffused with purplish gray in the temporal region. The iris is dark purplish brown with ivory gray flecks, and the pupil is vertically elliptic.

SEXUAL DIMORPHISM: We know of no external features that distinguish the two sexes; mean total lengths of the male and female are similar at La Selva (680 mm and 730 mm, respectively).

HABITAT: This arboreal, nocturnal species is often encountered in the laboratory clearing and at the River Station, but can also be found in areas of secondary forest.

DIET: Members of the genus *Sibon* have elongate teeth on the lower jaw that are designed for grasping slugs.

REPRODUCTION: This species is oviparous. Gravid females at La Selva have been recorded from mid-May (May 14, 1982) to late June (June 26, 1983). Two individuals laid four eggs each during late June and early July, 1983.

REMARKS: See remarks for the Ringed Slugeater *(S. annulatus)*.

TIGER RATSNAKE
Spilotes pullatus
Pl. 192

OTHER COMMON NAMES: Zumbadora, Mica
Gargantilla, Mica, Zopilota, Tropical Chicken
Snake, Tropical Ratsnake, Tiger Tree Snake
SIZE: Largest male 2,830 mm TL (n = 7);
largest female 2,430 mm TL (n = 6).

DISTRIBUTION: Caribbean slopes from San Luis Potosí, Mexico, to Argentina; Pacific slopes from Oaxaca, Mexico, to Ecuador.

IDENTIFYING FEATURES: This species is a large snake with a jet black dorsum offset with thin, diagonal, yellow bands anteriorly (uniform black posteriorly). The head is yellow with bold, black blotches; usually there is a broad, black band behind the eyes and black spots under the eyes. The only other large, shiny, black snake at La Selva is the Ebony Keelback *(Chironius grandisquamis)*, which has no bold, yellow markings.

SEXUAL DIMORPHISM: No external features distinguish the two sexes. The mean total lengths of the adult male and female are similar at La Selva (2,480 mm and 2,200 mm, respectively).

HABITAT: This snake is encountered most often crossing trails or climbing trees. It is diurnal and sometimes occurs in primary and secondary forest, but most of our records are from open areas, often near water. The species is arboreal but can be found on the ground, presumably traveling between trees.

DIET: This species is known to eat mammals, birds, bird eggs, frogs, and lizards (Alvarez del Toro 1983; Henderson and Hoevers 1977; Sexton and Heatwole 1965; Weyer 1990).

REPRODUCTION: This snake is oviparous, but nothing is known of the timing of egg production.

REMARKS: This species is aggressive when disturbed. It often holds its ground when approached by humans and is not afraid to display aggressively and strike. Aggressive displays usually involve a lateral compression of the neck region, hissing, and tail rattling. The tail of this snake is shorter at La Selva (25 percent of total length) than is described by Lee (1996) for this species in the Yucatan (30 to 35 percent of total length).

RETICULATED CROWNED SNAKE   *Tantilla reticulata*
Pl. 193

OTHER COMMON NAMES: Reticulated Centipede Snake

SIZE: Largest male 266 mm TL (n = 1); largest female 304 mm TL (n = 4).

DISTRIBUTION: Caribbean slopes from Nicaragua to Colombia; Pacific slopes from southwestern Costa Rica to Colombia.

IDENTIFYING FEATURES: This small littersnake has a middorsal, light brown stripe (one or two scale rows) and two pairs of light brown, dorsolateral stripes, each bordered below by dark brown (less than one scale row). The head is dark brown with a light yellow nuchal collar that is interrupted at the middorsal plane by a posterior extension of the dark head coloration. The supralabials are light anterior and posterior to the eye, but are dark below the eye and at the tip of the snout. The venter is yellow. This species has a more extensive light nuchal collar than the Orange-bellied Crowned Snake *(T. ruficeps)* and also differs from the latter species in having a yellow belly, rather than an orange or pink one. The Reticulated Crowned Snake differs from the Faded Dwarf Snake *(Trimetopon pliolepis)* in having a white or yellow (not tan) nuchal collar and from the Pink-bellied Littersnake *(Rhadinaea decorata),* the Orange-bellied Littersnake *(Urotheca guentheri),* and the Long-tailed Littersnake *(U. decipiens)* in lacking bold white or yellow stripes.

SEXUAL DIMORPHISM: No external features distinguish the two sexes.

HABITAT: The specimens that we have seen from La Selva were all found in or near clearings next to cacao *(Theobroma cacao)* groves.

DIET: The species is rear-fanged and is presumed to eat centipedes.

REPRODUCTION: Presumably this snake lays eggs.

ORANGE-BELLIED CROWNED SNAKE   *Tantilla ruficeps*
Pl. 194

SIZE: Largest male 391 mm TL (n = 4); largest female 334 mm TL (n = 2).

DISTRIBUTION: Caribbean slopes from Nicaragua to Costa Rica; Pacific slopes from Costa Rica to western Panama.

IDENTIFYING FEATURES: The dorsum of this small littersnake is dark brown with a thin, middorsal, black stripe (less than one scale row) from the neck to the tip of the tail. A pair of dorsolateral, tan stripes is present, each less than one scale row wide and bordered below by dark brown (one scale row). The head is dark brown to black. A light nuchal collar is present but is widely bisected dorsally, creating white cheek patches; the supralabials are white except for the posterior margin of supralabial number four and the ventral portions of supralabials five and six, which are dark brown. The venter is white anteriorly shading to orange and/or pink posteriorly. The most similar other species is the Reticulated Crowned Snake *(T. reticulata)*, which has a nearly complete nuchal collar (failing to meet only at the middorsal plane) and a white or yellow belly. The Orange-bellied Crowned Snake can be distinguished from the Pink-bellied Littersnake *(Rhadinaea decorata)* and the Orange-bellied Littersnake *(Urotheca guentheri)* by the presence of white cheek patches and from the Faded Dwarf Snake *(Trimetopon pliolepis)* by white rather than tan markings in the nuchal region.

SEXUAL DIMORPHISM: No external features distinguish the two sexes. Mean total length is similar between the adult male and female at La Selva (328 mm and 318 mm, respectively).

HABITAT: We have seen five specimens from La Selva, all of which were found in secondary forest.

DIET: The species is rear-fanged and is presumed to eat centipedes.

REPRODUCTION: This snake is presumed to lay eggs.

REMARKS: The specific epithet used here is that recommended by Savage (2002); older literature refers to this snake as *T. melanocephala.*

TRICOLORED CROWNED SNAKE   *Tantilla supracincta*
Pl. 195

OTHER COMMON NAMES: Coral Falsa

SIZE: We have no data for this species from La Selva; elsewhere in Costa Rica it reaches sizes of ca. 370 mm TL.

DISTRIBUTION: Caribbean slopes from southern Nicaragua to central Panama; Pacific slopes from the Osa Peninsula, Costa Rica, to Ecuador.

IDENTIFYING FEATURES: Like all members of the genus, this species is small in size. It mimics tricolored coralsnakes but differs from them in having bands rather than rings. The tip of the snout (rostral scale) is black, followed by a light (white or yellow) band across the snout in front of the eyes. The top of the head has a wide, black band to the posterior portion of the parietal scales. A light patch occurs on the side of the face behind each eye. The body consists of a repeated sequence of bands of the following colors: red (seven to 15 scale rows wide), black (two scale rows wide), light (white or yellow; one scale row wide), and black (two scale rows wide). This is the tricolored dyad pattern of Savage and Slowinski (1992). Typically, the snake has 10 to 11 red bands. The belly is immaculate white or cream. The dorsal bands often are offset, failing to meet middorsally. The species is rear fanged, with the enlarged teeth attached to venom glands. This species can be distinguished from the two true tricolored coralsnakes at La Selva, Allen's Coralsnake *(Micrurus alleni)* and the Central American Coralsnake *(M. nigrocinctus),* because the latter two have red rings that contact yellow or white ones, rather than black ones. This same feature also distinguishes the Tricolored Crowned Snake from the three other tricolored colubrids at La Selva, the Central American Coralsnake Mimic *(Erythrolamprus mimus),* the Skinkeater *(Scaphiodontophis annulatus),* and the Calico Snake *(Oxyrhopus petolarius).* Finally, the Tropical Milksnake *(Lampropeltis triangulum)* differs from the Tricolored Crowned Snake in being larger, having rings instead of bands, and having more than 20 red rings.

SEXUAL DIMORPHISM: No external features distinguish the two sexes.

HABITAT: This species is a member of an assemblage of small snakes that live in leaf litter. It is rare at La Selva, and little is known of its habitat preferences.

DIET: Nothing is known of the diet of this species.

REPRODUCTION: This snake is presumed to lay eggs.

REMARKS: This species is known at La Selva from a single individual palped from a juvenile Clelia *(Clelia clelia)* and photographed by E.D. Brodie, III (1990, personal communication). We follow Wilson (1987) for the specific epithet; older literature refers to the species as *T. annulata.*

ORANGE-BELLIED SWAMP SNAKE
Pl. 196

Tretanorhinus
nigroluteus

OTHER COMMON NAMES: Orangebelly Swamp Snake

SIZE: We have no measurements of this species from La Selva; elsewhere it reaches sizes of ca. 900 mm TL (Wilson and Meyer 1985).

DISTRIBUTION: Caribbean slopes from Veracruz, Mexico, to Colombia; Pacific slopes of Honduras and Nicaragua.

IDENTIFYING FEATURES: This species of snake is characterized by an extremely elongate head with a prominent black, postorbital stripe. The body is dark brown with a series of paired, indistinct, small (one to three scales), black spots along the entire length of the dorsum. The dark spots are sometimes offset slightly and may coalesce to form a middorsal zigzag pattern. A distinct, light, lateral stripe, more or less one scale row wide, extends from the angle of the jaw along the side of the body to about one-third the length of the snake. The light stripe then converges with the light ventral coloration. A dark brown, ventrolateral stripe transverses below the light stripe beginning at the infralabials and continuing posteriorly the length of the body. The venter is light brown to yellow with a few amorphous, dark, midventral blotches.

SEXUAL DIMORPHISM: We know of no external features that distinguish the two sexes.

HABITAT: This snake prefers slow-moving or standing water with thick aquatic vegetation. The species is rare at La Selva.

DIET: This animal eats tadpoles, frogs, and fish (Lee 1996; Scott 1969).

REPRODUCTION: This species lays eggs, but nothing is known of the timing of reproduction or of clutch size.

REMARKS: To our knowledge, this aquatic snake is known from La Selva from one museum specimen (CRE 70; a tail taken from a crab and identified by N. J. Scott, Jr.) and from an individual that was captured in an aquatic trap (E.D. Brodie III 1990, personal communication).

FADED DWARF SNAKE   *Trimetopon pliolepis*
Pl. 197

SIZE: Largest male 437 mm TL (n = 2); largest female 515 mm TL (n = 2).

DISTRIBUTION: Caribbean slopes of Costa Rica; Pacific slopes from central Costa Rica to Panama.

IDENTIFYING FEATURES: This species is one of many nondescript, small, brown littersnakes. The dorsum is dark grayish brown with faint, lateral, light stripes. The head is brown with a lighter brown nuchal collar. The supralabials are light and the ventral scutes are uniform white or yellow with large, black spots on the lateral-most part of each scute. This species can be distinguished from the Reticulated Crowned Snake (*Tantilla reticulata*) and the Orange-bellied Crowned Snake (*T. ruficeps*) by the presence of a tan, rather than white, nuchal collar.

Additionally, the possession of a nuchal collar distinguishes this species from the Pink-bellied Littersnake *(Rhadinaea decorata)*, and the absence of bold, light, lateral stripes distinguishes it from the Orange-bellied Littersnake *(Urotheca guentheri)* and the Long-tailed Littersnake *(U. decipiens)*.

SEXUAL DIMORPHISM: No external features distinguish the two sexes. Our small sample from La Selva does not indicate any difference in the mean total lengths of the adult male and female (429 mm and 467 mm, respectively).

HABITAT: All individuals that we have observed were active during the day and were captured crossing trails in primary and secondary forest.

DIET: This snake is rear-fanged and eats small lizards and frogs. One specimen at La Selva contained an adult Litter Gecko *(Lepidoblepharis xanthostigma)*.

REPRODUCTION: This species is presumed to lay eggs.

REMARKS: A second species from this genus, Viquez's Dwarf Snake *(Trimetopon viquezi)*, is known only from the type specimen, which was collected at Siquirres, Costa Rica. At least one Faded Dwarf Snake from our sample of La Selva had paired prefrontal scales, a characteristic of Viquez's Dwarf Snake and not the Faded Dwarf Snake. However, this individual did not have a red belly, a feature of Viquez's Dwarf Snake. Any member of this genus captured at La Selva should be compared carefully with the description of Viquez's Dwarf Snake (see under "Additional Species").

LONG-TAILED LITTERSNAKE   *Urotheca decipiens*
Pl. 198

SIZE: Largest individual 305 mm TL (n = 1; but with a broken tail).

DISTRIBUTION: Caribbean slopes of Costa Rica and Panama; Pacific slopes from central Costa Rica to Colombia.

IDENTIFYING FEATURES: This small littersnake has a dorsal ground color of deep chocolate brown. Two ventrolateral, light stripes extend from the neck to the tip of the tail on each side of the body; the dorsal-most stripe is very thin and tan, and the ventral-most one is white bordered below by black. The head is brown, followed by a wide, orange, pink, or yellow nuchal collar that includes the posterior three-fourths of the parietal shields and the anterior two to three dorsal scale rows. The supralabials are white with a dark spot in the center of each, and the venter is white. The iris is dark to reddish brown. This species differs from the Pink-bellied Littersnake *(Rhadinaea decorata)* and the Orange-bellied Littersnake *(U. guentheri)* by having a bold, light nuchal collar and differs from the Reticulated Crowned Snake *(Tantilla reticulata),* the Orange-bellied Crowned Snake *(T. ruficeps),* and the Faded Dwarf Snake *(Trimetopon pliolepis)* by possessing bold, light, lateral stripes.

SEXUAL DIMORPHISM: We know of no external features that distinguish the two sexes.

HABITAT: This is a rare member of the assemblage of small snakes that live in leaf litter. It has been taken from primary and cacao *(Theobroma cacao)* forests at La Selva, but little is known of its habitat preferences.

DIET: Like its congeners, this species is presumed to eat frogs.

REPRODUCTION: This snake presumably lays eggs.

REMARKS: Our generic designation follows that of Savage and Crother (1989); previous authors refer to this snake as *Rhadinaea decipiens.*

HALLOWEEN SNAKE
PI. 199

Urotheca euryzona

OTHER COMMON NAMES: Coral Falsa, Black Halloween Snake

SIZE: Largest male 805 mm TL (n = 5); largest female 785 mm TL (n = 7).

DISTRIBUTION: Caribbean slopes from Nicaragua to Panama; Pacific slopes from Panama to Ecuador.

IDENTIFYING FEATURES: This medium-sized snake is a bicolored coralsnake mimic. The color pattern is ringed (the dark markings completely encircle the body), and the head cap is black except for a small, white spot on the anterior supralabials, below the nares. The light rings are thin (usually two scale rows) and are orange dorsally, changing abruptly to white at the ventral-most two scale rows and continuing onto the ventral scutes; the light rings number 30 to 40 along the length of the body. The dark rings are much wider, 10 to 18 scale rows each. Thus, the pattern is orange and black dorsally and white and black ventrally; the colors are vivid because the scales are smooth and shiny. This color pattern conforms to the TZ pattern of Savage and Slowinski (1992). The only similar species are the Many-banded Coralsnake *(Micrurus mipartitus)*—which has light rings (orange, white, or pink) that are uniform in color and are much more numerous (more than 45)—and the young Calico Snake *(Oxyrhopus petolarius)*, which is banded instead of ringed, so the belly is uniformly light.

SEXUAL DIMORPHISM: No external features distinguish the two sexes. However, the male averages a slightly greater mean total length than the female at La Selva (716 mm and 598 mm, respectively).

HABITAT: This species has been found active day and night in both primary and secondary forest. However, the species is more active at night, when it can be found on the ground. It has been observed especially often around the Cantarana swamp.

DIET: This species primarily eats frogs. Diet records from La Selva include the Broad-headed Rainfrog *(Eleutherodactylus megacephalus)*, the Northern Masked Rainfrog *(E. mimus)*, and the Reticulated Sheepfrog *(Gastrophryne pictiventris)*.

REPRODUCTION: This species is presumed to lay eggs.

REMARKS: The unbroken tail in this species is unusually long and tapers to a fine point. However, the tail breaks easily and, therefore, is often lost, leaving a blunt tip. We have seen the tail writhe like that of a lizard when shed. Although this snake does not possess enlarged rear fangs, the bite is known to cause severe reactions in humans (Seib 1980). We follow the generic designation of Savage and Crother (1989); previous authors refer to this snake as *Pliocercus euryzonus.* When handled, this snake may twist and twitch with rapid side-to-side motions.

ORANGE-BELLIED LITTERSNAKE   *Urotheca guentheri*
Pl. 200

SIZE: Largest male 385 mm TL (n = 2); largest female 520 mm TL (n = 3).

DISTRIBUTION: Caribbean slopes from Honduras to Panama; Pacific slopes from the Osa Peninsula, Costa Rica, to Panama.

IDENTIFYING FEATURES: The dorsum of this small littersnake is deep chocolate brown with a faint, dark gray brown, middorsal stripe (three scale rows wide) bordered by dark reddish brown (two scale rows). Two light, dorsolateral stripes (one scale row wide) extend from the neck to the tip of the tail on each side of the body. The more ventral light stripe is bordered above and below by black (one scale row). The head is dark brown with a pair of off-white spots on the anterolateral aspect of the parietal shield. The supralabials are white and confluent with the more ventral light stripe. Three white spots, aligned with the middorsal and dorsolateral stripes, are present on the neck. The venter is salmon. This species is distinguished from the Long-tailed Litter-snake *(U. decipiens)* by the light spot behind the eye (on the anterolateral aspect of the parietal scale) and from the Pink-bellied Littersnake *(Rhadinaea decorata)* by the presence of bold, light stripes along the entire length of the body. The Reticulated Crowned Snake *(Tantilla reticulata)*, the Orange-bellied Crowned Snake *(T. ruficeps)*, and the Faded Dwarf Snake *(Trimetopon plio-lepis)* all have light nuchal collars or cheek patches that are not present in the Orange-bellied Littersnake.

SEXUAL DIMORPHISM: We know of no external features that distinguish the two sexes.

HABITAT: This species is typically seen on trails in primary forest but slithers off when approached.

DIET: This species is rear-fanged and, like many other leaf litter-snakes, eats frogs.

REPRODUCTION: This snake is oviparous. We palpated three eggs in a female from La Selva (June 16, 1983).

REMARKS: The rear fangs are associated with venom glands. The Orange-bellied Littersnake has an unusually long tail that is often lost. Our generic designation follows that of Savage and Crother (1989); previous authors refer to this snake as *Rhadinaea guen-theri.*

FALSE TERCIOPELO   *Xenodon rabdocephalus*
Pl. 201
OTHER COMMON NAMES: False Fer-de-Lance
SIZE: Largest male 680 mm TL (n = 6);
largest female 810 mm TL (n = 10).
DISTRIBUTION: Caribbean slopes from Veracruz,
Mexico, to Brazil; Pacific slopes from Guerrero,
Mexico, to Ecuador.

IDENTIFYING FEATURES: This is the only Terciopelo *(Bothrops as-per)* mimic at La Selva, but it is a good one. The snake is medium in size, and the dorsal pattern consists of a series of grayish brown blotches, each with a light border (one scale row wide). These blotches become narrow along the sides and interdigitate with dark gray, hourglass-shaped blotches that are bordered by black, and that are wide along the sides and narrow at the middorsum. The top of the head is dark brown with light flecks, and the head coloration may extend posteriorly as a large, heart-shaped blotch on the nape of the neck. A V-shaped, light, ocular stripe extends from a point on top of the snout anterior to the eyes, over the top of each eye, to the angle of the jaw; this stripe is bordered by dark brown. The venter is peach with numerous black punctations. The iris is round and light in color, a feature that can be noted from a safe observation distance.

SEXUAL DIMORPHISM: We know of no external features that distinguish the two sexes. The mean total length of the adult male at La Selva is shorter (474 mm) than the female (695 mm).

HABITAT: The species is terrestrial and diurnal and can be found in primary and secondary forests. Scott (1969) lists it as inhabiting riparian zones.

DIET: This species is rear-fanged and is a diurnal predator of frogs and toads. Two individuals at La Selva were known to have consumed the Broad-headed Rainfrog *(Eleutherodactylus megacephalus)*.

REPRODUCTION: This species is presumed to be oviparous.

REMARKS: When approached, this snake freezes and flares the quadrate and squamosal bones, causing the head to expand and assume a triangular shape like that of a pitviper. If further disturbed, it thrashes about as described by Lee (1996). We have not observed the voluntary bleeding at the mouth described by Swanson (1945).

Elapidae

This family contains venomous snakes that have a pair of short, immobile fangs located at the front of the mouth. The approximately 280 species within this family are found on all major continents except Antarctica. Most elapids are terrestrial, however one radiation contains aquatic species that have invaded the marine environment. New World representatives of this family are called coralsnakes; most have bright colors designed to warn potential predators that they are venomous. Three field characteristics help in the identification of coralsnakes. First, the iris is black, and the eye is small and located within a black head cap. Therefore, from a safe distance, no eye is easily visible. Second, unlike some colubrid mimics, the light and dark markings completely encircle the body as rings, a feature that can be determined by carefully examining the lateral surface of the snake or by lifting it with a stick. Third, the head of a true coralsnake is small and attaches to the body in a way that leaves no distinct neck region. Three species of coralsnakes are present at La Selva, and no additional species are known from other localities in the Caribbean lowland wet forests of Costa Rica.

Key to the Elapidae at La Selva

1a   Body bicolored throughout length (red, orange, or white, and black; select this choice for forms where the dorsal half of the light band is orange and the ventral half is white or yellowish) . . . . *Micrurus mipartitus*

1b   Body tricolored (red or orange, black, and yellow or white)
. . . . 2

     2a   Black head cap extending posteriorly as a lanceolate projection . . . . *Micrurus alleni*

     2b   Black head cap more or less straight, or U-shaped along posterior border . . . . *Micrurus nigrocinctus*

ALLEN'S CORALSNAKE   *Micrurus alleni*
Pl. 202
OTHER COMMON NAMES: Coral, Coral Macho, Coralillo
SIZE: Largest male 561 mm TL (n = 1). Elsewhere this species reaches sizes of up to ca. 1,070 mm TL (Campbell and Lamar 1989).
DISTRIBUTION: Caribbean slopes from Honduras to western Panama; Pacific slopes from central Costa Rica to Panama.
IDENTIFYING FEATURES: This midsized snake possesses a repeated series of rings with a black-yellow-red-yellow pattern (the tricolored monad of Savage and Slowinski [1992]). The red rings are the widest (about 12 to 14 scale rows wide), the yellow rings are the narrowest (about two to three scale rows wide), and the black rings are intermediate (six to eight scale rows wide). Approximately 17 red rings occur along the length of the body. The anterior portion of the head has a black skullcap from the tip of the

snout to the eyes; this cap extends as a lanceolate shape posteriorly to the back of the head, bisecting the first yellow ring. The color pattern and general body size and shape of Allen's Coralsnake are most similar to the Central American Coralsnake *(M. nigrocinctus)*. The major difference between the two species is that the black skullcap of the Central American Coralsnake has a straight border with the first yellow ring, which is a wide structure covering the back of the head. Five species of rear-fanged colubrids mimic the tricolored coralsnakes. In two of these, the Tropical Milksnake *(Lampropeltis triangulum)* and the Tricolored Crowned Snake *(Tantilla supracincta)*, the red marks are bordered by black ones. Of the remaining mimics, one species, the Calico Snake *(Oxyrhopus petolarius)*, has more red bands (more than 20) and two species, the Central American Coralsnake Mimic *(Erythrolamprus mimus)* and the Skinkeater *(Scaphiodontophis annulatus)*, have fewer red bands (less than 15) than Allen's Coralsnake. In addition, these three mimics usually have markings that are offset at the middorsal line. Finally, the small, black eye hidden in the black skullcap of Allen's Coralsnake is a good field characteristic for distinguishing it from the mimics.

SEXUAL DIMORPHISM: No external features distinguish the two sexes.

HABITAT: This snake is rare at La Selva. Habitat and activity are presumed to be similar to the Central American Coralsnake.

DIET: This species is known to eat swamp eels (genus *Synbranchus*) and lizards (Roze 1996), but presumably consumes other snakes as well.

REPRODUCTION: The female is oviparous, but nothing is known of the timing of reproduction.

MANY-BANDED CORALSNAKE   *Micrurus mipartitus*
Pl. 203

OTHER COMMON NAMES: Gargantilla, Coral, Coralillo, Red-tailed Coralsnake, Bicolored Coral Snake

SIZE: We have no measurements of this species from La Selva; elsewhere this snake reaches sizes of up to ca. 1,130 mm TL (Campbell and Lamar 1989).

DISTRIBUTION: Caribbean slopes from central Nicaragua to Panama; Pacific slopes from eastern Panama to Ecuador.

IDENTIFYING FEATURES: This is midsized species is the only bicolored coralsnake at La Selva. It usually has black and orange rings, but the latter may be faded to shades of yellow, pink, or white. This snake has 46 to 56 light rings, each of which is two to four scale rows wide. The black rings are six to eight scale rows wide. The head is black from the tip of the snout to the level of the eyes. A white to yellowish white ring covers the rest of the head. Two species of colubrids at La Selva, the Calico Snake *(Oxyrhopus petolarius)* and the Halloween Snake *(Urotheca euryzona)*, mimic this species. Both of these have fewer than 35 light or red bands. Additionally, the Calico Snake has bands (color not extending onto the belly scutes) rather than rings, and the Halloween Snake has light rings that are two-toned (i.e., orange dorsally, white ventrally). In a few individuals of the Many-banded Coralsnake, the light band is orange at the center, fading to yellow or white laterally. This creates a tricolored morph (Leenders et al. 1996). However, this morph still has more than 40 light rings, whereas the other tricolored coralsnakes and colubrid mimics at La Selva have fewer than 30 light rings.

SEXUAL DIMORPHISM: Savage and Vial (1974) suggest that the male has more rings than the female.

HABITAT: This species is rarely encountered, largely because of its secretive habits. Habitat and activity are presumed to be similar to the Central American Coralsnake *(M. nigrocinctus)*.

DIET: This snake is presumed to eat other snakes.

REPRODUCTION: This species is presumed to lay eggs.

REMARKS: Some recent works refer to this species as *M. nigrofasciata*, restricting *M. mipartitus* to South American forms (Campbell and Lamar 1989; Roze 1984).

CENTRAL AMERICAN CORALSNAKE
Pl. 204

Micrurus nigrocinctus

OTHER COMMON NAMES: Coral, Coralillo, Coral Macho

SIZE: Largest male 610 mm TL (n = 4); largest female 890 mm (n = 3).

DISTRIBUTION: Caribbean slopes from Belize to northern Colombia; Pacific slopes from Oaxaca, Mexico, to northern Colombia.

IDENTIFYING FEATURES: This is the most frequently seen coralsnake at La Selva. It is medium-sized (up to approximately 1 m), and the color pattern commences with a black skullcap covering the front of the head and a yellow ring covering the back of the head. This is followed by a repeated series of black, yellow, red, and yellow rings; the red ring is the widest (about eight scale rows wide), the yellow ring is the narrowest (about two scale rows wide, except on the tail), and the black ring is intermediate (about five scale rows wide). This snake has 17 to 18 red rings along the length of the body. This pattern conforms to the tricolored monad of Savage and Slowinski (1992). The black skullcap and first yellow ring usually have a straight border between them, unlike Allen's Coralsnake *(M. alleni),* in which a lanceolate projection of the black skullcap bisects the first yellow ring. Of the five species of rear-fanged colubrids at La Selva that mimic tricolored coralsnakes, two species, the Tropical Milksnake *(Lampropeltis triangulum)* and the Tricolored Crowned Snake *(Tantilla supracincta),* have red markings that are bordered by black ones. One of the remaining mimics, the Calico Snake *(Oxyrhopus petolarius),* has more than 20 red bands, and the other two mimics, the Central American Coralsnake Mimic *(Erythrolamprus mimus)* and the Skinkeater *(Scaphiodontophis annulatus)* have fewer than 15 red rings or bands. In addition, these three mimics usually have markings that are offset at the middorsal line, whereas those of the Central American Coralsnake almost always are symmetrical across the body. Finally, the small, black eye hidden in the black skullcap of the Central American Coralsnake is a good field characteristic for distinguishing it from the mimics, which typically have large, easily visible eyes.

SEXUAL DIMORPHISM: No external features distinguish the two sexes. However, our data suggest that the female may reach larger sizes than the male.

HABITAT: The Central American Coralsnake is active day or night but is seen most frequently at dusk. It is observed fairly commonly moving around the buildings of the laboratory clearing.

DIET: This species eats small to medium-sized snakes and, occasionally, caecilians, lizards, and lizard and snake eggs (Greene and Seib 1983; Roze 1996; Seib 1985b; Smith and Grant 1958). One individual at La Selva consumed an adult Tropical Night Lizard *(Lepidophyma flavimaculatum).*

REPRODUCTION: This species is oviparous, but little is known of its patterns of egg production.

REMARKS: The False Tree Coral *(Rhinobothryum bovalli),* a species known from the Caribbean lowlands of Costa Rica but not yet known from La Selva, mimics the Central American Coralsnake. The False Tree Coral is likely to key to the Central American Coralsnake in our key, but does not fit our description of that species. Any coral-patterned snake with keeled scales should be compared with the description of the False Tree Coral (see under "Additional Species").

CROCODYLIA

The order Crocodylia is an ancient lineage of reptiles that includes only 21 living species. These animals have large bodies as adults, elongate, flat heads, and laterally compressed tails. Although of limited species richness, the order is widespread, occurring on all continents except Antarctica.

All crocodilians have internal fertilization and lay shelled eggs. Internal fertilization occurs via the insertion of a male's penis into the cloaca of a female. This process is preceded by vocal, visual, and chemical signals given by the male in order to attract a female. Both sexes participate in courtship displays that contribute to the formation of a pair-bond: the male and female take turns riding on each other's back. Additionally, the male rubs secretions from his chin glands onto the top of the head and neck of the female as part of the courtship ritual. Through participating in these behaviors, females actively select from among potential mates. Once inseminated, either the female or both sexes create a nest. Such nests can be composed of mounds of sand, or may include vegetation and soil piled on top of the eggs to help incubate them. One or both parents defend the nest until the eggs hatch, and offspring frequently stay near one or both parents during the early stages of life. Vocalizations are given between parents and offspring during this period of parental care. In all of these respects, reproduction in crocodilians has features similar to that observed in ground-nesting birds.

As adults, crocodilians often are top predators in the aquatic habitats that they occupy. They eat a wide variety of prey, including items as small as aquatic snails and as large as ungulates (cloven-hoofed mammals such as deer or goats) and, occasionally, humans. Crocodilians are thought to be keystone species in certain environments because they regulate the abundance of organisms that might otherwise dominate the fauna and because, during times of drought, they create depressions in soil that retain water, thereby allowing other aquatic organisms to survive through this period of environmental stress. Crocodilian eggs and juveniles incur heavy mortality from predatory mammals, birds, large lizards, and other crocodilians. Adults, however, have few predators except for humans.

Crocodilians are found in stagnant and slow-moving water. They frequently bask and are best observed from boats traveling along major rivers. At night, smaller individuals can be found by wading in small streams and searching the banks for the characteristic eye shine of crocodilians. Because all individuals can inflict painful bites, we do not recommend that they be captured.

Key to the Crocodilians of La Selva

1a   Snout short and wide . . . . Alligatoridae
1b   Snout long and narrow . . . . Crocodylidae

Alligatoridae

This family contains the alligators and caimans, which are characterized by the presence of an enlarged tooth at the anterior end of the lower jaw that fits into a socket in the upper jaw and is therefore visible along the side of the upper jaw. Thus, the snout is short and wide. Only one species within this family is present at La Selva, and it is the only species that occurs in the Caribbean lowlands of Costa Rica.

SPECTACLED CAIMAN
Pls. 205, 206
Caiman crocodylus

OTHER COMMON NAMES: Caimán

SIZE: No measurements are available for individuals from La Selva; elsewhere the adult male can reach sizes of 2,000 mm TL and the female 1,200 mm TL (Alvarez del Toro 1974).

DISTRIBUTION: Caribbean coastal drainages from Honduras to the Amazon Basin; Pacific coastal drainages from southern Mexico to Ecuador.

IDENTIFYING FEATURES: This is the smaller of the two crocodilians found at La Selva. A distinct sharp ridge between the eyes is characteristic of the species and is the source of the English common name. The dorsum is uniform dark gray, brown, or black in the adult. In the juvenile, the color pattern consists of bold, brownish yellow and dark (brown or black) bars across the dorsum and the sides of the body and tail. This species can be distinguished from the larger American Crocodile *(Crocodylus acutus)* by its smaller size, broader snout, and darker skin, and by the sharp ridge between the eyes.

SEXUAL DIMORPHISM: The adult male is much larger and has a bigger head and longer tail than the female (Alvarez del Toro 1974; Thorbjarnarson 1994).

HABITAT: Large adult Spectacled Caiman often bask along the Rio Puerto Viejo; they are often seen from bridges or boats, or by people floating down the river. Also, the Spectacled Caiman can be seen at the Research and Cantarana swamps or can be found by walking the major streams at night and searching the water surface for red or orange eye shine. Elsewhere, this species tends to bask during the late morning and late afternoon hours (Thorbjarnarson 1995).

DIET: The juvenile Spectacled Caiman eats aquatic insects, snails, and crustaceans. As the animal grows larger, vertebrate prey—principally fish and mammals—are added to the diet. The adult eats mostly vertebrate prey during the dry season and invertebrates during the wet season. The animal captures vertebrate prey by floating slowly and snapping the head rapidly sideways, or lying on the bottom of a water column and jumping (usually through a school of fish), or lying perpendicular to a riffle zone and snapping at prey as they swim by the animal's head (Alvarez del Toro 1974; Thorbjarnarson 1993a, 1993b).

REPRODUCTION: Breeding occurs during the late dry season or early wet season. The male attracts a female by giving a call described as a dull roar. The nest is composed of vegetation and soil scraped into a mound along a riverbank, often in forested areas. The female deposits 20 to 30 eggs in the nest, with clutch size increasing with increasing female size. Both parents attend the nest and attack intruders aggressively. The offspring vocalize while hatching, and the parents excavate the nest in response to this cue. The parents may also roll unhatched eggs in their mouths to help the hatching process. The offspring stay with the female parent for up to one month before dispersing singly or in groups (Alvarez del Toro 1974; Thorbjarnarson 1994).

REMARKS: Although this species is hunted in the region for its hide and as a source of meat, La Selva seems to have maintained a healthy population of Spectacled Caiman over the years. However, no study of its status or life history has been performed at this site.

Crocodylidae

This family is characterized by the presence of an enlarged tooth located at the anterior end of the lower jaw and visible along the margin of the upper jaw when the mouth is shut. The snout is long and thin. La Selva has but a single species of this family, and it is the only species known to occur in the Caribbean lowland wet forests of Costa Rica.

AMERICAN CROCODILE   *Crocodylus acutus*
Pl. 207

OTHER COMMON NAMES: Cocodril, Lagarto
SIZE: We have no measurements of this species from La Selva; elsewhere this species reaches sizes of ca. 7,300 mm TL, but most are 3,000 to 4,000 mm TL (Levy 1991).
DISTRIBUTION: Caribbean coastal areas from southern Florida to Venezuela (including much of the Antilles); Pacific coastal areas from Sinaloa, Mexico, to Peru.
IDENTIFYING FEATURES: The adult of this species is unmistakable, given its large size. The body color of the adult is tan to olive; the juvenile is more boldly marked with light olive and black bars along the dorsum and the sides of the body and tail. The American Crocodile can be distinguished from the Spectacled Caiman by its relatively long and narrow snout, its olive green coloring, and the rounded bump (not a sharp ridge) between the eyes.
SEXUAL DIMORPHISM: The male is larger than the female and possesses bulkier jaw musculature.
HABITAT: This species prefers large, slow-moving rivers, swamps, lagoons, and estuaries. It has recently been observed most frequently in the Rio Sarapiquí (at trail marker STR 2300), where it can be seen basking on large sand bars during the day or may be found by searching the water surface at night for eye shine. It has been observed less frequently in the Rio Puerto Viejo.
DIET: Foraging studies have not been performed at La Selva. In Mexico, the diet of the juvenile is composed principally of aquatic insects but also includes small frogs, turtles, birds, and mammals; the adult primarily forages on fish but may eat a variety of other vertebrates (Alvarez del Toro 1974).

REPRODUCTION: Data from Mexico indicate that nesting occurs late in the dry season (Alvarez del Toro 1974). The female creates a mound on a sandy beach, deposits from 30 to 70 eggs in the nest, covers them with sand and leaf litter, and then attends the nest until hatching. The female is thought to assist in the hatching process and to provide care for the offspring for an unknown period of time after birth.

REMARKS: Given the known historical range of the species, the American Crocodile was undoubtedly present at La Selva in recent times. However, it was missing from La Selva for a number of years, probably as a result of hunting pressure. The skin of this animal is valued in the garment trade, leading to current conservation concerns for the entire order (Groombridge 1982). The species is noted elsewhere for its nasty temperament, so those swimming in or floating down the river should take warning.

ADDITIONAL SPECIES

IN THIS SECTION, we provide species accounts for amphibians and reptiles that are known from the lowland wet forests of eastern Costa Rica but are not known from La Selva. Although some of these taxa are distributed near to La Selva and, therefore, are likely to be discovered at the site in the future, others are unlikely to occur at La Selva because their distributional limits do not approach the site. The latter are included for those visiting other sites in the lowland wet forests of Costa Rica.

AMPHIBIA

Gymnophiona
Caeciliidae

LA LOMA CAECILIAN   *Dermophis parviceps*

REMARKS: The La Loma Caecilian is the most widespread caecilian in Costa Rica and is known from lowland wet forest sites. It has a pinkish head, a purplish slate body, and a gray belly. The adult is up to 217 mm TL. This species is similar in color and body shape to the Purple Caecilian *(Gymnopis multiplicata)*. To distinguish between the two, the position of the tentacle must be determined. In the La Loma Caecilian the tentacle is approximately midway between the eye and the external naris; in the Purple Caecilian this structure is located immediately anterior to the eye. The tentacle is a small structure that requires a hand lens to locate. The La Loma Caecilian is a burrowing species that may be found on the surface of the ground after intense rains or under rotting logs.

Caudata
Plethodontidae

ALVARADO'S SALAMANDER   *Bolitoglossa alvaradoi*

OTHER COMMON NAMES: Moravia de Chirripo Salamander

REMARKS: This species, like all members of the genus, has a robust body with four well-developed legs. Unlike any other members of the genus *Bolitoglossa* in Costa Rican lowland wet forests, this species has a dorsum colored with large, light blotches on a black background. The light blotches often fuse along the sides of the body. This species is large (up to 159 mm SVL; Leenders 2001) and arboreal. It is likely to key to the Ridge-headed Salamander

(B. colonnea) in our key but differs from that species (and the Striated Salamander *[B. striatula]*) in having toes that extend beyond the webbing.

STRIATED SALAMANDER   *Bolitoglossa striatula*
OTHER COMMON NAMES: Cukra Mushroomtongue Salamander
REMARKS: This species is characterized by a uniform cream, tan, brick red, or yellow dorsal coloration with several irregular, longitudinal, brown stripes. It also has toes that are completely encased in webbing. The species is medium-sized, reaching lengths of up to 140 mm SVL (Leenders 2001), and is most frequently encountered in disturbed areas such as abandoned pastureland. Both characteristics separate this species from Alvarado's Salamander *(B. alvaradoi)*. The Striated Salamander can be distinguished from the Ridge-headed Salamander *(B. colonnea)* because it is more colorful and lacks the projecting ridge between the eyes of the latter species.

Anura
Leptodactylidae

WARTY RAINFROG   *Eleutherodactylus bufoniformis*
OTHER COMMON NAMES: Rusty Robber Frog
REMARKS: This species of the genus *Eleutherodactylus* has a broad head, toes that lack expanded disks, and a dorsum covered by whitish bumps. Additionally, the top of the head has a series of distinct crests. The species is terrestrial and is most similar to the Broad-headed Rainfrog *(E. megacephalus)*. However, the cranial crests and white tubercules distinguish the Warty Rainfrog from the Broad-headed Rainfrog. The Warty Rainfrog is found in Costa Rica only in the extreme southeastern corner, near the Panama border.

GAIGE'S RAINFROG   *Eleutherodactylus gaigeae*
OTHER COMMON NAMES: Fort Randolph Robber Frog
REMARKS: This species is readily distinguishable from all other

members of the genus because of its distinctive color pattern. The dorsum is black with a lateral, red stripe along each side. This color pattern appears to mimic that of the Striped Dart-poison Frog *(Phyllobates lugubris);* however, the lateral stripe of the Striped Dart-poison Frog is yellow, not red. In Costa Rica, Gaige's Rainfrog is known only from the extreme southeastern corner of the country near the Panama border.

SOUTHERN MASKED RAINFROG   *Eleutherodactylus gollmeri*

OTHER COMMON NAMES: Evergreen Robber Frog, Southern Mimicking Rainfrog

REMARKS: This species is similar in color pattern and shape, and likely is closely related to, the Northern Masked Rainfrog *(E. mimus)*. However, the two can be distinguished from each other based on the amount of webbing between the toes of the hind foot and the general body color. The Southern Masked Rainfrog has limited webbing on the hind feet, less than one-third the length of the toes, whereas the webbing is greater than one-third the length of the toes in the Northern Masked Rainfrog; and the Southern Masked Rainfrog is tan in color, whereas the Northern Masked Rainfrog is gray. The Southern Masked Rainfrog is found in the southeastern corner of Costa Rica and appears to be replaced by the Northern Masked Rainfrog at more northerly sites in the Caribbean lowlands.

TURBO WHITE-LIPPED FOAMFROG   *Leptodactylus poecilochilus*

OTHER COMMON NAMES: Turbo White-lipped Frog, White-lipped Frog

REMARKS: This species was recently documented from the La Guaria annex at La Selva. It is much more common on the Pacific slopes of Costa Rica and is known from only a few localities in Atlantic drainages (Savage 2002). Its discovery at La Selva may indicate habitat alterations that have allowed this and other species (see species accounts for Striped Basilisk *[Basiliscus vittatus]* and Yellow-headed Gecko *[Gonatodes albogularis]*) to invade the area surrounding La Selva. It is a small frog that breeds in roadside pools and grassy swamps and is similar in size and shape to the

Fringe-toed Foamfrog *(Leptodactylus melanonotus)*. However, the Turbo White-lipped Foamfrog has a dark gray upper lip above which is a bold, white stripe extending from the tip of the snout, under the tympanum, to the axilla. Additionally, this species has a bold, white stripe extending along the length of the back of the thigh. The Fringe-toed Foamfrog has a series of bold, white spots on the lower lip, but no distinctive markings on the upper lip or the back of the thigh. The male Fringe-toed Foamfrog develops two black spines at the base of the thumb during the breeding season, but the male Turbo White-lipped Foamfrog develops no such spines. In our key, the Turbo White-lipped Foamfrog fails couplet 14, where the thigh coloration fits neither choice.

Hylidae

SPURRELL'S LEAF FROG   *Agalychnis spurrelli*
OTHER COMMON NAMES: Gliding Leaf Frog
REMARKS: This species is similar to the Red-eyed Leaf Frog *(A. callidryas)* in size and general color pattern; however, the flanks of Spurrell's Leaf Frog are orange rather than blue with yellow bars. In Costa Rica Spurrell's Leaf Frog is found on the Osa Peninsula and in the extreme southeastern corner of the country near the Panama border.

HIGHLAND FRINGE-LIMBED TREEFROG   *Hyla miliaria*
OTHER COMMON NAMES: Cope's Brown Treefrog
REMARKS: This is one of two "flying" treefrogs found in Costa Rica. Like all members of the family, this species has enlarged toe pads that adhere to leaves. The Highland Fringe-limbed Treefrog has extensive webbing between all fingers and toes. Additionally, there is a fleshy fringe of skin along the posteroventral margin of the arms and legs. These features can be used to create an airfoil when the species leaps from arboreal perches, allowing it to parachute to a lower perch. No other hylid frog of the lowland wet forests of Costa Rica has such extensive webbing and a fringed lateral margin on the limbs. The adult can be as large as or larger than the species in the genus *Agalychnis*. The male produces a loud growl when calling.

Centrolenidae

GHOST GLASSFROG   *Centrolenella ilex*
OTHER COMMON NAMES: Limon Giant Glassfrog
REMARKS: The Ghost Glassfrog is known from Rara Avis, a field station upslope from La Selva. Little is known of its distribution and habits; however, it is larger in size than other glassfrogs in Costa Rica (up to 37 mm snout-to-vent; Leenders 2001) and has eyes that are positioned so that they point directly forward. Both features should readily distinguish this species from other wet forest glassfrogs. The iris of the eye is silver with a black reticulum, and the bones are green. The species is most similar to the Powdered Glassfrog *(Hyalinobatrachium pulveratum)* in that it is green with large white spots and has green bones. However, the Ghost Glassfrog differs from the Powdered Glassfrog in having a white peritoneal sheath that covers the internal organs, whereas the peritoneum is clear and colorless in the Powdered Glassfrog.

Ranidae

TAYLOR'S LEOPARD FROG   *Rana taylori*
OTHER COMMON NAMES: Peralta Frog
REMARKS: Like all ranid frogs, this species has extensive webbing between the toes but no expanded toe pads. This species differs from all other ranids of the Atlantic lowlands of Costa Rica by possessing large, rounded, dark spots on a tan or gray ground color. The species also has a pair of dorsolateral ridges that are white or light yellow. The call of this species is a complicated series of chortles and squalls.

REPTILIA

Testudines
Emydidae

ORNATE SLIDER   *Trachemys ornata*
REMARKS: This aquatic turtle is found in the slow-moving waters and estuaries of the Atlantic coast of Costa Rica. It has not yet been documented from La Selva, but is likely to be in both the Rios Sarapiquí and Puerto Viejo. It is most similar to the Black Wood Turtle *(Rhinoclemmys funerea),* but differs from that species in having bold, yellow stripes on the face and neck and an oblong red mark on the posterolateral part of the head. No other turtle has this pattern of head coloration. Also there are dark spots on the marginal scutes.

Squamata
Polychrotidae

PUERTO RICAN CRESTED ANOLE   *Ctenonotus cristatellus*
REMARKS: This anole was introduced to the city park in Limón, Costa Rica. From this site, it has expanded its range northward and southward along the coast. Because it thrives in areas disturbed by humans, it may eventually reach La Selva via range expansion. The species is grayish to dark brown with indistinct darker blotching. The male possesses an impressive crest that can be erected along the back of the head, neck, middorsum, and tail. The male also has a burnt orange dewlap. This species is most similar to the Lemur Anole *(Norops lemurinus)* but differs from that species in the presence of an extensive caudal crest in the male and in the color of the dewlap in the male (red in the Lemur Anole).

Teiidae

FOUR-LINED WHIPTAIL   *Ameiva quadrilineata*

REMARKS: This species of the genus *Ameiva* is widespread in disturbed areas of the Atlantic coast of Costa Rica. It is quite similar to the Central American Whiptail *(A. festiva)* in body shape and color pattern. However, the Four-lined Whiptail lacks a middorsal, light stripe. Instead, the Four-lined Whiptail has a pair of light stripes along the lateral portion of the body. The adult male has a bright blue head, in contrast to the tan head of the female.

Anomalepidae

PINK-HEADED BLINDSNAKE   *Helminthophis frontalis*

REMARKS: This snake is unmistakable because of its blunt, essentially eyeless head and its short, thin body. The mouth of this snake is difficult to see and opens only wide enough to ingest small insects (ants and termites). The body is dark gray with a light pink head and neck. Because this animal appears to have no head, it is frequently mistaken for an invertebrate.

Boidae

RAINBOW BOA   *Epicrates cenchria*

REMARKS: This snake is similar to the Annulated Tree Boa *(Corallus annulatus)* but differs from that species in being largely terrestrial (rather than arboreal), having a thick, muscular neck (rather than a thin one), and having light ocelli (faded and edged in dark) down the middle of the back (rather than along the sides). Also, the Rainbow Boa has smooth, shiny scales that produce a shiny iridescence. The Rainbow Boa is known from habitat similar to that near La Selva and, therefore, is expected to occur at that site.

Colubridae

DIAMONDBACK RACER — *Drymobius rhombifer*

REMARKS: The Diamondback Racer is a brown snake with a series of large, rhomboid blotches extending down the middle of the back. This is a terrestrial species, but it may roost in shrubs at night. The species is similar in size and color pattern to the Northern Cat-eyed Snake *(Leptodeira septentrionalis)*. However, the Diamondback Racer differs from that species in having larger blotches and a round pupil (rather than a vertical one).

FALSE TREE CORAL — *Rhinobothryum bovalli*

REMARKS: The False Tree Coral is an arboreal coralsnake mimic. Its pattern of colors is most similar to the Central American Coralsnake *(Micrurus nigrocinctus)*. However, the False Tree Coral has keeled scales, a feature that makes the scales look rough; the Central American Coralsnake has smooth, shiny scales. Additionally, the False Tree Coral has a large head with a distinct neck. The head of the Central American Coralsnake is small and rounded, and the neck is barely, if at all, differentiated from the head. The head cap of the False Tree Coral is black, but the enlarged scales on the head are edged in red and/or yellow, making the head scales and eye obvious. In the Central American Coralsnake, the head cap is solid black, making it difficult to distinguish the small black eye or individual head scales. Finally, in the False Tree Coral, the black ring is the widest, with red rings being intermediate in size. In the Central American Coralsnake, the opposite pattern occurs. Probably because of its arboreal habits, the False Tree Coral is known from only a few localities. Virtually nothing is known of its life history.

DEGENHARDT'S SCORPION-EATER — *Stenorrhina degenhardtii*

REMARKS: This is a small, leaf litter species that lives up to its common name by eating principally scorpions. The adult is either uniform brown or light brown with a series of dark brown blotches. This species is most similar in size and color pattern to the two species of the genus *Leptodeira* at La Selva, the Southern

Cat-eyed Snake *(L. annulata)* and the Northern Cat-eyed Snake *(L. septentrionalis)*. However, Degenhardt's Scorpion-eater differs from those species in being terrestrial (rather than arboreal), diurnal (rather than nocturnal), and stout in body shape (rather than thin).

VIQUEZ'S DWARF SNAKE   *Trimetopon viquezi*

REMARKS: This species is known only from the type specimen, which was collected at Siquirres in the Atlantic lowlands of Costa Rica. Therefore, it may be expected at La Selva. The key characteristics for this species are the presence of paired prefrontal scales (single in the Faded Dwarf Snake *[T. pliolepis]*), 17 dorsal scale rows, seven supralabials, two postocular scales, fewer than 40 subcaudal scales, and a bright red venter. Because few specimens of the genus *Trimetopon* have been collected, any individual captured should be examined carefully for these characteristics.

FIELD DATA SHEET FOR AMPHIBIANS AND REPTILES

GYMNOPHIONA

Total length: _____ Mass: _____

Position of tentacle: Closer to eye or naris?_____

Color notes:

CAUDATA

Total length: _____ Mass: _____

Length of tail: ☐ 1 × body length
                 ☐ 2 × body length   ☐ 3 × body length

Color notes:

ANURA

Total length: _____ Mass: _____

Bone color: ☐ Green   ☐ White

Toe pads: ☐ Absent from all digits
             ☐ Present on outer two digits of hands
             ☐ Present on all digits of hand and rounded
             ☐ Present on all digits of hand and pointed

Hind foot webbing:
        ☐ Extending half way or more along length of toes
        ☐ Absent or extending less than half way along toes

Color notes:

CROCODYLIA

Total length: _____ Mass: _____

Snout shape: ☐ Lower tooth visible along outside of jaw
☐ Lower tooth not visible along outside of jaw

Color notes:

TESTUDINES

Total length: _____ Mass: _____

Carapace shape: ☐ Flattened ☐ Domed

Plastron: ☐ 12 scutes ☐ 10–11 scutes
☐ Reduced and cross-shaped

Color notes:

SQUAMATA

Total length: _____ Mass: _____

Dorsal scales: ☐ Shiny ☐ Velvety
☐ Beaded ☐ Keeled

Ventral scales: ☐ Same size as dorsum
☐ Rectangular ☐ Wide and flat

Head shape: ☐ With crests ☐ With spines
☐ No crest or spines

Pupil shape: ☐ Vertical ☐ Round

Tail length: ☐ About as long as body
☐ Much longer than body

Tail shape: ☐ Round ☐ Laterally compressed
☐ Dorsoventrally flattened

Color notes:

GLOSSARY

Acarines Members of the order Acarina; mites and their kin.

Adpressed Describing placement of the limbs against the body such that the hand is as far posterior and the foot as far anterior as possible on one side of the body.

Alluvial Of or pertaining to soils deposited by settling of sediments transported in water.

Amniotes Land vertebrates that produce amniotic eggs; reptiles, birds, and mammals.

Amplectant Of or pertaining to being in amplexus.

Amplexus The position adopted by frogs during reproduction; this typically involves the male grasping the female from above and behind so that the male's arms grasp the axilla or groin of the female.

Anal Of or pertaining to the anus or cloaca.

Anastomosing Being interconnected or weblike.

Anterior Toward the head of an organism.

Anterolateral Toward the head and side of an organism.

Anticoagulant A substance that stops or retards coagulation of blood.

Antillean Of the Antilles islands in the Caribbean Ocean.

Apical pits Small depressions on the tips of dorsal scales of some snakes.

Aposematic Of a bright and/or warning color.

Apterygotes Members of the order Apterygota; collembolans and their kin.

Aquatic Of the water.

Araneids Members of the order Araneae; web-building spiders and their kin.

Arboreal Living in trees and/or shrubs.

Arthropods Members of the phylum Arthropoda; insects, crustaceans, spiders, and their kin.

Axil The base of a leaf.

Axilla The armpit.

Band A rectangular marking of color, oriented across the back of the body but not extending onto the belly.

Bask To rest in sunlight in order to gain heat.

Biparental Describing activities involving both male and female parents.

Bipedal Describing locomotion propelled by the two hind limbs and with the forelimbs off of the ground.

Blattids Members of the order Blattoidea; roaches and their kin.

Bromeliads Plants of the family Bromeliacaeae; air plants and their kin.

Buttress A thin, wide, winglike projection from the trunks of some canopy trees.

Caecilians The common name for members of the order Gymnophiona.

Calcar A flaplike extension of skin found on the heel of some frogs.

Canopy The region of the forest associated with the tops of trees.

Carapace The upper portion of the shell of a turtle.

Carboniferous The geological time period from 290 to 360 million years before present.

Caribbean Plate A piece of lithosphere containing most of the Greater Antillean islands and moving as a unit via tectonic activities.

Carnivory The consumption of vertebrates.

Caudal fin The flattened portion of the tail used by fish and larval amphibians to swim.

Chorioallantoic placenta A structure designed for transmission of nutrients from an adult female to a developing offspring through the chorion and allantois of the embryo; this structure is associated with live birth and is the type of placenta found in eutherian mammals.

Cloaca The common opening of the reproductive, digestive, and urinary tracts.

Clutch An aggregation of eggs deposited by a single female during a single reproductive event.

Cocos Plate A piece of lithosphere containing Isla del Cocos and moving as a unit via tectonic activities.

Coleopterans Members of the order Coleoptera; beetles and their kin.

Colubrids Members of the family Colubridae.

Concave Bowed inward.

Congener A member of the same genus.

Conspecific A member of the same species.

Copulation The transmission of sperm to a female's reproductive tract via insertion of a specialized intromittent organ by a male.

Cornified skin The thickened (and frequently darkened) skin caused by deposition of keratin.

Costal grooves Depressions along the side of the body of a salamander that correspond to the ribs.

Crepuscular Active at dawn and/or dusk.

Cruciform In the shape of a cross.

Crustaceans Members of the order Crustacea; shrimp, crayfish, crabs, and their kin.

Cryptic Hidden from view; difficult to distinguish from background colors.

Cycloid Describing a scale that is flat, shiny, and with a half-moon-shaped posterior border.

Dermopterans Members of the order Dermoptera; earwigs and their kin.

Desiccation The process of drying.

Dewlap The laterally compressed flap of skin under the chin of some lizards that can be extended during behavioral interactions.

Dimorphism The condition of having two distinct body forms.

Diplopods Members of the order Diplopoda; millipedes and their kin.

Dipterans Members of the order Diptera; flies and their kin.

Direct development A developmental sequence that takes place within the egg in which the larval stage is skipped and the juvenile emerges as a small version of the adult.

Disks Padlike structures at the tips of the digits of some frogs, designed to adhere to smooth surfaces.

Diurnal Active during daylight hours.

Dorsal Toward the back of an organism.

Dorsolateral Toward the back but along the sides of an organism.

Dorsoventral A plane through the middle of the body separating the dorsal and ventral halves.

Dorsum The back of an organism.

Ectothermic Of or pertaining to gathering body heat from the environment.

Endemic Restricted to a specific locale.

Endotherms Animals that generate body heat through metabolic processes (mammals and birds).

Eocene A geological Epoch from 54 to 38 million years before present.

Epidermal scales Keratinized structures composing the skin of reptiles.

Epidermis The outermost layer of skin.

Epiphytes Plants that grow on other plants (e.g., orchids, bromeliads). Also called *epiphytic plants*.

External Toward the outside.

External nares Openings on the outside of the snout that lead into the nasal cavities.

Extinction The loss of all individuals of a species.

Extirpation The loss of all individuals of a population.

Fauna All of the animal species known from a region.

Femur The large, single bone of the thigh.

Finca The Spanish word for farm.

Flank The fleshy part of the side of the body between the ribs and the hind limbs.

Formicids Members of the family Formicidae; ants.

Gap An opening in the forest canopy created by fallen trees or branches.

Gap specialist An organism found only in treefall gaps.

Gastropods Members of the order Gastropoda; snails and slugs and their kin.

Generic designation The genus to which a species belongs.

Globose Round or globelike in shape.

Gonochoristic Having separate sexes (male and female) that do not change throughout an individual's lifetime.

Gravid Describing the reproductive condition in which a female has eggs in her oviducts.

Groin The area where the thigh meets the flank.

Guanine A nitrogenous base (from nucleic acids) that can form a sheath around internal organs in some frogs.

Heliothermic Describing an organism that positions itself in full sun to increase its body temperature.

Hemipenes The paired copulatory organs of male lizards and snakes.

Hemipterans Members of the order Hemiptera; true bugs and their kin.

Herbivory The consumption of plant material.

Herpetofauna The amphibian and reptilian species known from a particular region.

Homopterans Members of the order Homoptera; treehoppers and their kin.

Humeral hooks Bony structures present at the base of the humerus in some frogs.

Humerus The large, single bone of the upper arm.

Hymenopterans Members of the order Hymenoptera; ants, bees, wasps, and their kin.

Hyoid bones Small bones of the throat that allow tongue movement and suspend the larynx.

Iguanians A group containing the most recent common ancestor of iguanas and chameleons and all descendants of that ancestor; a group of families of squamates that use their tongues in prey acquisition.

Immaculate Without spots.

Infralabials The scales of the lower lip.

Insectivory The consumption of insects.

Intercalary cartilage The cartilage located between the last bony element of a digit and the toe pad in some frogs.

Interorbital Between the eyes.

Interscale epidermis The skin between the epidermal scales in reptiles (observable by stretching the skin).

Invagination An inward pocketing of the skin.

Invertebrates Animals without backbones.

Iris The tissue of the eye that regulates the amount of light entering the pupil.

Isopods Members of the class Isopoda; sowbugs, pillbugs, rolypolies, and their kin.

Isopterans Members of the order Isoptera; termites and their kin.

Isthmus of Tehuantepec The narrowest region in southern mainland Mexico.

Keeled Describing an epidermal scale with a pronounced central ridge.

Keratinized Composed of the protein keratin.

Key A tool used to identify species through a series of dichotomous choices.

Kinetic skull A skull in which the bones of both the upper and lower jaw move during feeding.

Labials The scales lining the lips.

Lamellae A series of wide, flat scales covering the bottoms of the toes of some lizards.

Lanceolate In the shape of an arrow.

Larvae Individuals in the sexually immature, aquatic stage of development for amphibians.

Lateral Toward the sides of the body.

Leaf litter The dead leaves and sticks that accumulate on the forest floor.

Lepidopterans Members of the order Lepidoptera; butterflies and moths and their kin.

Lichenose Colored with speckles of green, white, tan, and/or yellow so that the color pattern resembles a lichen.

Loreal The region between the nostril and eye on the side of the face; of or pertaining to a specific scale in this area in some lizards and snakes (fig. 13a).

Marginal scutes Scales found on the perimeter of the carapace of a turtle.

Maxilla The largest tooth-bearing bone of the upper jaw.

Medial Toward the midline of the body.

Median At the midline of the body.

Metamorphosis The process by which amphibian larvae transform to become juveniles.

Microhabitat The immediate space around an organism.

Middorsal Pertaining to the middle of the back.

Midventral Pertaining to the middle of the belly.

Miocene A geological Epoch from 24 to 5 million years before present.

Monophyletic Describing a taxon in which all of the organisms are derived from a common ancestor and that includes all descendants of that ancestor.

Morphological Referring to external form or shape.

Naris The external opening to the nose; nostril. The plural is *nares.*

Nazca Plate A piece of lithosphere containing the Galapagos Islands and moving as a unit during tectonic activity.

Neonate A newborn organism.

Nocturnal Active at night.

North American Plate A piece of lithosphere containing North America and moving as a unit via plate tectonics.

Nuchal Referring to the dorsal surface of the neck.

Ocelli Eye-shaped spots; color markings consisting of a light center surrounded by a dark border.

Oligochaetes Members of the class Oligochaeta; earthworms.

Omnivory The consumption of plant and animal materials.

Opposable Describing digits whose tips are capable of being brought together and are capable of grabbing.

Oral Referring to the mouth.

Orbits The spaces in the skull where the eyes are located.

Orthopterans Members of the order Orthoptera; grasshoppers, katydids, crickets, and their kin.

Oviduct The tube that conducts eggs from the ovary to the uterus.

Oviparous Describing organisms that lay eggs.

Paedomorphosis The retention of juvenile characteristics in a sexually mature individual.

Palp To cause regurgitation of a prey item by manual manipulation.

Palpate To manually manipulate an organism in order to determine the presence of internal objects (e.g., the number of eggs in the uterus).

Panamanian Portal The gap that existed between lower Central America and South America prior to the Pleistocene.

Paravertebral Referring to structures that are next to the middorsal line.

Parietal Referring to scales or bones located behind the eyes, on the top of the head.

Parietal shields Large, paired scales on the top of the head posterior to the eyes of squamates.

Parotoid gland A large gland found on the head immediately behind the eyes of toads.

Parthenogenesis Development of an ovum without fertilization by a sperm.

Pectoral amplexus A behavior in anurans in which the male grasps the female specifically in the axillary region during mating.

Pectoral girdle The bony structure that functions to attach the arm to the body; the shoulder. Also called *shoulder girdle.*

Pelvic amplexus A behavior in anurans in which the male grasps the female specifically in the groin region during mating.

Pelvic girdle The bony structure that functions to attach the hind leg to the body; the hip.

Pericardium The connective tissue that surrounds the heart.

Peripheral Around the outer edges.

Peritoneal sheath The lining of the abdominal cavity.

Phallodeum A specialized copulatory structure of male caecilians.

Phylogenetic Of or pertaining to evolutionary relationships.

Placental mammals Members of the subclass Theria; mammals in which fertilized eggs implant on the uterine wall, where development of offspring occurs.

Plastron The lower portion of the shell of a turtle.

Pliocene A geological Epoch from 5 to 1.8 million years before present.

Poeciliids Members of the family Poeciliidae; guppies and their kin.

Ponerine ants Members of the subfamily Ponerinae; large black ants including balas, or bullet ants *(Paraponera clavipes)*, and their kin.

Postanal Behind the anus.

Posterior Toward the tail.

Posteroventral Toward the tail and the belly.

Postorbital Behind the eye.

Predaceous Describing organisms that are predators; capable of consuming animals.

Prehensile tail A tail with the ability to grasp structures.

Premaxilla Anteriormost tooth-bearing bone of the upper jaw.

Prepollical At the base of the first finger or thumb.

Primary forest An old-growth forest; forest that has not been modified by human activities.

Pupil The opening that allows light to enter the eye.

Quadrate The bone in the upper jaw that articulates with the lower jaw in amphibians and reptiles.

Radiation A group of species that are all descended from a single common ancestor.

Rear-fanged Having enlarged teeth at the back of the maxilla; these often are associated with specialized glands that produce venoms.

Reticulum A netlike pattern.

Ring A colored portion of the body that completely encircles the body.

Riparian Next to streams or rivers.

Scutes Large, platelike scales on the shell of a turtle or the belly of a snake.

Second growth forest A forest that has been cut and has regenerated.

Semifossorial Burrowing in superficial layers of soil or under leaf litter.

Serrate Having a sawlike edge.

Shank The portion of the leg between the knee and the ankle.

Size dimorphism A condition in which the sexes are of different sizes.

South American Plate A part of the lithosphere containing South America and moving as a unit via plate tectonics.

Specific epithet The second part of the Latin name of a species.

Spermatophore A packet of sperm covered by a gelatinous outer coat created by male salamanders.

Squamosal A bone at the back of the skull that articulates with the quadrate.

Subcanopy The forest vegetative zone immediately below the canopy.

Subocular Below the eye.

Suborbital Below the bony eye socket.

Subterranean Below the ground.

Superciliaries The scales above the eye.

Supralabials The scales of the upper lip.

Supraocular tubercle An enlarged conical bump above the eye.

SVL Snout-to-vent length; the body length of an animal measured from the tip of the snout to the opening of the cloaca.

Synonomies Duplicate scientific names for one species.

Syntopic Describing the condition of having several species in the same habitat.

Tadpole An individual of the larval phase of frogs and toads.

Tapetum lucidum The guanine-containing layer of the inner portion of the eye that is responsible for eye shine in some amphibians and reptiles.

Tarsus The ankle.

Teleost fishes Members of the subclass Teleostei; the major radiation of living bony fishes.

Temporal Relating to the region of the skull or head near the ear openings.

Tentacle A chemosensory and tactile structure of caecilians located between the nostril and eye.

Terrestrial Living on land or on the ground.

Tettegoniids Members of the family Tettigoniidae; long-horned grasshoppers, or katydids, and their kin.

TL Total length; the distance from the tip of the snout to the tip of the tail.

Tricolor dyad A color pattern of red bands or rings separated from one another by a sequence of alternating bands or rings of black-light-black; the light bands may be white or yellow.

Tricolor monad A color pattern of red bands or rings separated from one another by a sequence of alternating bands or rings of light-black-light; the light bands may be white or yellow.

Tubercles Rounded bumps on the skin.

Tympanum The membrane that separates the external ear from the middle ear.

TZ pattern A tricolor pattern in which dark rings or bands contact light rings or bands that are red dorsally and white or yellow ventrally.

Umbilical Referring to the cordlike structure that attaches a developing embryo to the placenta or egg membranes of reptiles, birds, and mammals.

Understory The bottommost vegetative zone in the forest; consists of short trees and shrubs.

Urostyle A bony projection located between the ilia of an anuran sacrum.

Venter The belly.

Ventral Toward the belly.

Ventrolateral Toward the belly and the sides of the body.

Vertebral Along the middorsal line.

Vertebrates Organisms that have backbones; a taxon containing the most recent common ancestor of lampreys and crocodilians and all descendants of that ancestor.

Viscera Internal organs; guts.

Viviparous Having a reproductive mode in which offspring develop in the uterine tract of the female, who gives birth to live young.

Vocal sac The inflatable throat pouch(es) of male frogs.

Vocal slit An opening in the floor of the mouth or throat used by male frogs to produce vocalizations.

Voucher A catalogued museum specimen that documents the occurrence of a species at a particular site at a particular time.

Yolking follicles Enlarged eggs in ovaries that are preparing to be ovulated.

LITERATURE CITED

Acuna, R.A. 1993. *Las tortugas continentales de Costa Rica.* San José, Costa Rica: Universidad de Costa Rica.

Allen, R., and W.T. Neill. 1956. Effect of marine toad toxins on man. *Herpetologica* 12:150–151.

Alvarez del Toro, M. 1960. *Reptiles de Chiapas.* Tuxtla Gutierrez, Mexico: Instituto Zoológico del Estado.

Alvarez del Toro, M. 1974. *Los Crocodylia de Mexico.* Mexico City: Instituto Mexicano de Recursos Naturales Renovables.

Alvarez del Toro, M. 1983. *Los reptiles de Chiapas,* 3rd ed. Tuxtla Gutierrez, Mexico: Instituto de Historia Natural.

Alvarez del Toro, M., and H.M. Smith. 1958. Notulae herpetologicae Chiapasiae II. *Herpetologica* 12:3–17.

Amaral, A. 1927. Studies of Neotropical Ophidia. VII. An interesting collection of snakes from west Colombia. *Bull. Antivenin Inst. Am.* 2:44–47.

Andrews, R.M. 1979. The lizard *Corytophanes cristatus:* An extreme sit-and-wait predator. *Biotropica* 11:136–139.

Andrews, R.M. 1983. *Norops polylepis.* In *A Costa Rican natural history,* edited by D.H. Janzen. Chicago: University of Chicago Press.

Andrews, R.M. 1991. Population stability of a tropical lizard. *Ecology* 72:1204–1217.

Backwell, P.R.Y., and M.D. Jennions. 1993. Mate choice in the neotropical frog, *Hyla ebraccata:* Sexual selection, mate recognition and signal selection. *Anim. Behav.* 45:1248–1250.

Baugh, J.R., and D.C. Forester. 1994. Prior residence effect in the dart-poison frog, *Dendrobates pumilio. Behaviour* 131:207–224.

Beebe, W. 1944. Field notes on the lizards of Kartabo, British Guiana, and Caripito, Venezuela. Part 2: Iguanidae. *Zoologica* 29:195–216.

Beuttell, K., and J.B. Losos. 1999. Ecological morphology of Caribbean anoles. *Herpetol. Monogr.* 13:1–28.

Bezy, R. L. 1989. Morphological differentiation in unisexual and bi-sexual xantusiid lizards of the genus *Lepidophyma* in Central America. *Herpetol. Monogr.* 3:61–80.

Blake, J. G., F. G. Stiles, and B. A. Loiselle. 1990. Birds of La Selva Bio-logical Station: Habitat use, trophic composition, and migrants. In *Four Neotropical rainforests*, edited by A. H. Gentry. New Haven, Conn.: Yale University Press.

Blaney, R. M., and P. K. Blaney. 1978. Additional specimens of *Amas-tridium veliferum* Cope (Serpentes: Colubridae) from Chiapas, Mexico. *Southwest. Nat.* 23:692.

Blankenship, E. L. 1993. The effects of diet on the predatory larva of the smoky jungle frog (Leptodactylidae: *Leptodactylus penta-dactylus*). Master's thesis, Auburn University, Alabama.

Bock, B. C. 1987. *Corytophanes cristatus:* Nesting. *Herpetol. Rev.* 18:35.

Bock, B. C., and A. S. Rand. 1989. Factors influencing nesting syn-chrony and hatching success at a green iguana nesting aggrega-tion in Panama. *Copeia* 1989:978–986.

Bogert, C. M., and A. P. Porter. 1966. The differential characteristics of the Mexican snake related to *Geophis dubius* (Peters). *Am. Mus. Novit.* 2277:1–19.

Boyer, D. M., G. M. Garrett, J. B. Murphy, H. M. Smith, and D. Chiszar. 1991. In the footsteps of Charles C. Carpenter: Facultative strike-induced chemosensory searching and trail-following behavior of bushmasters *(Lachesis muta)* at Dallas Zoo. *Herpetol. Monogr.* 9:161–168.

Braker, H. E., and H. W. Greene. 1994. Population biology: Life histo-ries, abundance, demography, and predator-prey interactions. In *La Selva: Ecology and natural history of a Neotropical rain forest,* edited by L. A. McDade, K. S. Bawa, H. A. Hespenheide, and G. S. Hartshorn. Chicago: University of Chicago Press.

Bridegam, A. S., C. M. Garrett, and D. T. Roberts. 1990. Cannibalism in two species of arboreal pitviper, *Trimeresurus wagleri* and *Bothriechis schlegeli. Herpetol. Rev.* 21:54–55.

Brodie, E. D., III. 1993. Differential avoidance of coral snake banded patterns by free-ranging avian predators in Costa Rica. *Evolution* 47:227–235.

Burge, R. M. 1995. An arboreal burrower: The dwarf boa. *Vivarium* 7:46–49.

Burghardt, G. M., H. W. Greene, and A. S. Rand. 1977. Social behavior in hatchling green iguanas: Life at a reptile rookery. *Science* 195: 689–691.

Buttenhoff, P.A., and R.C. Vogt. 1995. *Bothrops asper* (Nauyaca): Cannibalism. *Herpetol. Rev.* 26:146–147.

Buttenhoff, P.A., and R.C. Vogt. 1997. *Bothrops asper* (nauyaca, sorda). In *Historia natural de Los Tuxtlas,* edited by E. González Soriano, R. Dirzo, and R.C. Vogt. Mexico City: Universidad Nacional Autónoma de México.

Caldwell, J.P. 1994. Natural history and survival of eggs and early larval stages of *Agalychnis calcarifer* (Anura: Hylidae). *Herpetol. Nat. Hist.* 2:57–66.

Campbell, H.W. 1973. Ecological observations on *Anolis lionotus* and *Anolis poecilopus* (Reptilia: Sauria) in Panama. *Am. Mus. Novit.* 2516:1–29.

Campbell, J.A., and W.W. Lamar. 1989. Venomous reptiles of Latin America. Ithaca, N.Y.: Comstock Publishing Associates.

Campbell, J.A., L.S. Ford, and J.P. Karges. 1983. Resurrection of *Geophis anocularis* Dunn with comments on its relationships and natural history. *Trans. Kans. Acad. Sci.* 86:38–47.

Carr, A. 1954. The windward road. New York: A.A. Knopf.

Censky, E.J., and C.J. McCoy. 1988. Female reproductive cycles of five species of snakes (Reptilia: Colubridae) from the Yucatan Peninsula. *Biotropica* 20:326–333.

Clark, D.A. 1994. Plant demography. In *La Selva: Ecology and natural history of a Neotropical rain forest,* edited by L.A. McDade, K.S. Bawa, H.A. Hespenheide, and G.S. Hartshorn. Chicago: University of Chicago Press.

Clark, D.B. 1988. The search for solutions: Research and education at the La Selva Biological Station and their relation to ecodevelopment. In *Diversity and conservation of tropical rainforests,* edited by F. Almeda and C.M. Pringle. San Francisco: California Academy of Science.

Clark, D.B. 1990. La Selva Biological Station: A blueprint for stimulating tropical research. In *Four Neotropical rainforests,* edited by A.H. Gentry. New Haven, Conn.: Yale University Press.

Conners, J.S. 1989. *Oxybelis fulgidus* (green vine snake): Reproduction. *Herpetol. Rev.* 20:73.

Corn, M.J. 1974. Report on the first certain collection of *Ungaliophis panamensis* from Costa Rica. *Caribb. J. Sci.* 14:167–175.

Corn, M.J. 1981. Ecological separation of *Anolis* lizards in a Costa Rican rain forest. Ph.D. diss., University of Florida, Gainesville.

Crimmins, M.L. 1937. A case of *Oxybelis* poisoning in man. *Copeia* 1937:233.

Crother, B.I., J.A. Campbell, and D.M. Hillis. 1992. Phylogeny and historical biogeography of the palm-vipers, genus *Bothriechis:* Biochemical and morphological evidence. In *Biology of the pit-vipers,* edited by J.A. Campbell and E.D. Brodie, Jr. Tyler, Tex.: Selva Press.

Daly, J.W., J.M. Garraffo, T.F. Spande, C. Jaramillo, and A.S. Rand. 1994. Dietary source for skin alkaloids of poison frogs (Dendrobatidae)? *J. Chem. Ecol.* 20:943–955.

Dioscoro, S.R. 1952. Preliminary notes on the giant toad, *Bufo marinus* (Linn.), in the Philippine Islands. *Copeia* 1952:281–282.

Ditmars, R.L. 1939. Field book of North American snakes. Doran, N.Y.: Doubleday.

Dixon, J.R. 1980. The Neotropical colubrid snake genus *Liophis:* The generic concept. *Milwaukee Publ. Mus. Contrib. Biol. Geol.* 31: 1–40.

Dixon, J.R., J.A. Wiest, Jr., and J.M. Cei. 1993. Revision of the Neotropical snake genus *Chironius* Fitzinger (Serpentes: Colubridae). *Museo Regionale di Scienze Naturali Monografie* 13: 1–279.

Donnelly, M.A. 1989a. Demographic effects of reproductive resource supplementation in a territorial frog, *Dendrobates pumilio. Ecol. Monogr.* 59:207–221.

Donnelly, M.A. 1989b. Effects of reproductive resource supplementation on space-use patterns in *Dendrobates pumilio. Oecologia* 81:212–218.

Donnelly, M.A. 1989c. Reproductive phenology and age structure of *Dendrobates pumilio* in northeastern Costa Rica. *J. Herpetol.* 23: 362–367.

Donnelly, M.A. 1991. Feeding patterns of the strawberry poison frog, *Dendrobates pumilio* (Dendrobatidae). *Copeia* 1991:723–730.

Donnelly, M.A. 1994a. Amphibian diversity and natural history. In *La Selva: Ecology and natural history of a Neotropical rain forest,* edited by L.A. McDade, K.S. Bawa, H.A. Hespenheide, and G.S. Hartshorn. Chicago: University of Chicago Press.

Donnelly, M.A. 1994b. Appendix 5. In *La Selva: Ecology and natural history of a Neotropical rain forest,* edited by L.A. McDade, K.S. Bawa, H.A. Hespenheide, and G.S. Hartshorn. Chicago: University of Chicago Press.

Donnelly, M.A. 1999. Reproductive phenology of *Eleutherodactylus bransfordii* (Anura: Leptodactylidae) in northeastern Costa Rica. *J. Herpetol.* 33:624–631.

Donnelly, M.A., and C. Guyer 1994. Patterns of reproduction and habitat use in an assemblage of Neotropical hylid frogs. *Oecologia* 98:291–302.

Donnelly, M.A., R.O. de Sá, and C. Guyer. 1990. Description of the tadpoles of *Gastrophryne pictiventris* and *Nelsonophryne aterrima* (Anura: Microhylidae), with a review of morphological variation in free-swimming microhylid larvae. *Am. Mus. Novit.* 2976:1–19.

Donnelly, M.A., C. Guyer, and R.O. de Sá. 1990. The tadpole of a dart poison frog, *Phyllobates lugubris* (Anura: Dendrobatidae). *P. Biol. Soc. Wash.* 103:427–431.

Donnelly, M.A., C. Guyer, D.M. Krempels, and H. E. Braker. 1987. The tadpole of *Agalychnis calcarifer* (Anura: Hylidae). *Copeia* 1987:247–250.

Duellman, W.E. 1963. Amphibians and reptiles of the rainforests of southern El Petén, Guatemala. *Univ. Kans. Mus. Nat. Hist. Publ.* 15:205–249.

Duellman, W.E. 1967a. Courtship isolating mechanisms in Costa Rican frogs. *Herpetologica* 23:169–183.

Duellman, W.E. 1967b. Social organization in the mating calls of some Neotropical anurans. *Am. Midl. Nat.* 77:157–163.

Duellman, W.E. 1970. The hylid frogs of Middle America. *Mus. Nat. Hist. Univ. Kans. Monogr.,* no. 1.

Duellman, W.E. 1978. The biology of an equatorial herpetofauna in Amazonian Ecuador. *Univ. Kans. Mus. Nat. Hist. Misc. Publ.,* no. 65.

Duellman, W.E., and J.J. Wiens. 1992. The status of the hylid frog genus *Oloygon* and the recognition of *Scinax* Wagler, 1830. *Occas. Pap. Mus. Nat. Hist. Univ. Kans.* 151:1–23.

Dundee, H.A., and E.A. Liner. 1974. Eggs and hatchlings of the tree snake *Leptophis depressirostris* (Cope). *Brenesia* 3:11–14.

Dunn, E.R. 1931. The amphibians of Barro Colorado Island. *Occas. Pap. Boston Soc. Nat. Hist.* 38:111–130.

Eaton, T.H. 1941. Notes on the life history of *Dendrobates auruatus*. *Copeia* 1941:93–95.

Echelle, A.A., A.F. Echelle, and H.S. Fitch. 1971. A new anole from Costa Rica. *Herpetologica* 27:354–362.

Echternacht, A.C. 1983. *Ameiva* and *Cnemidophorus* (Chisbalas, macroteid lizards). In *Costa Rican natural history,* edited by D.H. Janzen. Chicago: University of Chicago Press.

Ernst, C.H., and R.W. Barbour. 1989. *Turtles of the world.* Washington, D.C.: Smithsonian Institution Press.

Estes, R., K. de Queiroz, and J. Gauthier. 1988. Phylogenetic relationships within Squamata. In *Phylogenetic relationships of the lizard families,* edited by R. Estes and G. Pregill. Stanford, Calif.: Stanford University Press.

Etheridge, R. 1967. Lizard caudal vertebrae. *Copeia* 1967:284–295.

Fauth, J.E., B.I. Crother, and J.B. Slowinski. 1989. Elevational patterns of species richness, evenness, and abundance of the Costa Rican leaf-litter herpetofauna. *Biotropica* 21:178–185.

Fitch, H.S. 1970. Reproductive cycles in lizards and snakes. *Univ. Kans. Mus. Nat. Hist. Misc. Publ.*, no. 52.

Fitch, H.S. 1973a. A field study of Costa Rican lizards. *Univ. Kans. Sci. Bull.* 50:39–126.

Fitch, H.S. 1973b. Population structure and survivorship in some Costa Rican lizards. *Occas. Pap. Mus. Nat. Hist. Univ. Kans.*, no. 18.

Fitch, H.S. 1975. Sympatry and interrelationships in Costa Rican anoles. *Occas. Pap. Mus. Nat. Hist. Univ. Kans.*, no. 40.

Fitch, H.S. 1976. Sexual size differences in the mainland anoles. *Occas. Pap. Mus. Nat. Hist. Univ. Kans.*, no. 50.

Fitch, H.S. 1982. Reproductive cycles in tropical reptiles. *Occas. Pap. Mus. Nat. Hist. Univ. Kans.*, no. 96.

Fitch, H.S., A.F. Echelle, and A.A. Echelle. 1976. Field observations on rare or little known mainland anoles. *Univ. Kans. Sci. Bull.* 51: 91–128.

Fleishman, L.J. 1985. Cryptic movement in the vine snake *Oxybelis aeneus. Copeia* 1985:242–244.

Ford, L.S., and D.C. Cannatella. 1993. The major clades of frogs. *Herpetol. Monogr.* 7:94–117.

Frost, D.R., and R. Etheridge. 1989. A phylogenetic analysis and taxonomy of iguanian lizards (Reptilia: Squamata). *Univ. Kans. Mus. Nat. Hist. Misc. Publ.*, no. 81.

Gaffney, E.S., and P.A. Meylan. 1988. A phylogeny of turtles. In *The phylogeny and classification of tetrapods.* Vol. 1, *Amphibians, reptiles, birds,* edited by M.J. Benton. Oxford: Clarendon Press.

Galatti, U. 1992. Population biology of the frog *Leptodactylus pentadactylus* in a Central Amazonian rainforest. *J. Herpetol.* 26:23–31.

Gans, C. 1974. *Biomechanics: An approach to vertebrate biology.* Philadelphia, Pa.: J.B. Lippencott.

Gerhardt, H.C. 1994. The evolution of vocalizations in frogs and toads. *Ann. Rev. Ecol. Syst.* 25:293–324.

Glasheen, J.W., and T.A. McMahon. 1996. Size-dependence of water-running ability in basilisk lizards *(Basiliscus basiliscus). J. Exp. Biol.* 199: 2611–2618.

Good, D.A., and D.B. Wake. 1997. Phylogenetic and taxonomic implications of protein variation in the Mesoamerican salamander genus *Oedipina* (Caudata: Plethodontidae). *Rev. Biol. Trop.* 45: 1185–1208.

Greding, E.J., Jr. 1972. Call specificity and hybrid compatibility between *Rana pipiens* and three other *Rana* species in Central America. *Copeia* 1972:383–385.

Greene, H.W. 1969. Reproduction in a Middle American skink, *Leiolopisma cherrei* (Cope). *Herpetologica* 25:55–56.

Greene, H.W. 1975. Ecological observations on the red coffee snake, *Ninia sebae,* in southern Veracruz, Mexico. *Am. Midl. Nat.* 93: 478–484.

Greene, H.W. 1983. *Boa constrictor* (boa, bequer, boa constrictor). In *Costa Rican natural history,* edited by D.H. Janzen. Chicago: University of Chicago Press.

Greene, H.W. 1986. Diet and arboreality in the emerald monitor, *Varanus prasinus,* with comments on the study of adaptation. *Fieldiana Zool.* 31:1–12.

Greene, H.W. 1988. Species richness in tropical predators. In *Diversity and conservation of tropical rainforests,* edited by F. Almeda and C.M. Pringle. San Francisco: California Academy of Science.

Greene, H.W. 1992. The ecological and behavioral context for pitviper evolution. In *Biology of the pitvipers,* edited by J.A. Campbell and E.D. Brodie, Jr. Tyler, Tex.: Selva Press.

Greene, H.W. 1997. *Snakes: The evolution of mystery in nature.* Berkeley, Calif.: University of California Press.

Greene, H.W., and D.L. Hardy. 1989. Natural death associated with skeletal injury in the terciopelo, *Bothrops asper* (Viperidae). *Copeia* 1989:1036–1037.

Greene, H.W., and M.A. Santana. 1983. Field studies of hunting behavior by bushmasters. *Am. Zool.* 23:897.

Greene, H.W., and R.L. Seib. 1983. *Micrurus nigrocinctus* (coral, coral snake, coralillo). In *Costa Rican natural history,* edited by D.H. Janzen. Chicago: University of Chicago Press.

Groombridge, B. 1982. *The IUCN Amphibia-Reptilia red book.* Part 1, *Testudines, Crocodylia, Rhynchocephalia.* Gland, Switzerland: IUCN–The World Conservation Union.

Guyer, C. 1986. Seasonal patterns of reproduction of *Norops humilis* (Sauria: Iguanidae) in Costa Rica. *Rev. Biol. Trop.* 34:247–251.

Guyer, C. 1988a. Food supplementation in a tropical mainland anole, *Norops humilis:* Demographic effects. *Ecology* 69:350–361.

Guyer, C. 1988b. Food supplementation in a tropical mainland anole, *Norops humilis:* Effects on individuals. *Ecology* 69:362–369.

Guyer, C. 1990. The herpetofauna of La Selva, Costa Rica. In *Four Neotropical rainforests,* edited by A.H. Gentry. New Haven, Conn.: Yale University Press.

Guyer, C. 1994a. Appendix 6. In *La Selva: Ecology and natural history of a Neotropical rain forest*, edited by L.A. McDade, K.S. Bawa, H.A. Hespenheide, and G.S. Hartshorn. Chicago: University of Chicago Press.

Guyer, C. 1994b. The reptile fauna: Diversity and ecology. In *La Selva: Ecology and natural history of a Neotropical rain forest*, edited by L.A. McDade, K.S. Bawa, H.A. Hespenheide, and G.S. Hartshorn. Chicago: University of Chicago Press.

Guyer, C., and M.A. Donnelly. 1990. Length-mass relationships among an assemblage of tropical snakes in Costa Rica. *J. Trop. Ecol.* 6:65–76.

Guyer, C., and J.M. Savage. 1986. Cladistic relationships among anoles (Sauria: Iguanidae). *Syst. Zool.* 35:509–531.

Guyer, C., and J.M. Savage. 1992. Anole systematics revisited. *Syst. Biol.* 41:89–110.

Haines, T.P. 1940. Delayed fertilization in *Leptodeira annulata polysticta*. *Copeia* 1940:116–118.

Hallinan, T. 1920. Notes on lizards of the Canal Zone, Isthmus of Panama. *Copeia* 83:45–49.

Hardy, D.L. 1994. Snakebite and field biologists in Mexico and Central America: Report on ten cases with recommendations for field management. *Herpetol. Nat. Hist.* 2:67–82.

Hartshorn, G.H. 1983. Plants. In *Costa Rican natural history*, edited by D.H. Janzen. Chicago: University of Chicago Press.

Hartshorn, G.S., and B.E. Hammel. 1994. Vegetation types and floristic patterns. In *La Selva: Ecology and natural history of a Neotropical rain forest*, edited by L.A. McDade, K.S. Bawa, H.A. Hespenheide, and G.S. Hartshorn. Chicago: University of Chicago Press.

Hayes, M.P. 1991. A study of clutch attendance in the Neotropical frog *Centrolenella fleischmanni* (Anura: Centrolenidae). Ph.D. diss., University of Miami.

Hayes, M.P., J.A. Pounds, and W.W. Timmerman. 1986. An annotated list and guide to the amphibians and reptiles of Monteverde, Costa Rica. *Soc. Study Amph. Rept. Herptol. Circ.,* no. 17.

Heinen, J.T. 1992. Comparisons of community characteristics of the herpetofauna of leaf litter in abandoned cacao plantations and primary forest in a lowland tropical rainforest: Some implications for fauna restoration. *Biotropica* 24:420–430.

Henderson, R.W. 1974. Aspects of the ecology of the juvenile common iguana. *Herpetologica* 30:327.

Henderson, R.W. 1982. Trophic relationships and foraging strategies of some New World tree snakes *(Leptophis, Oxybelis, Uromacer)*. *Amphibia-Reptilia* 3:71–80.

Henderson, R.W. 1984. *Scaphiodontophis* (Serpentes: Colubridae): Natural history and test of a mimicry-related hypothesis. *Univ. Kans. Mus. Nat. Hist. Spec. Publ.* 10:185–197.

Henderson, R.W., and M.H. Binder. 1980. The ecology and behavior of vine snakes *(Ahaetulla, Oxybelis, Thelotornis, Uromacer):* A review. *Milwaukee Publ. Mus. Contrib. Biol. Geol.* 37:1–38.

Henderson, R.W., and L.G. Hoevers. 1977. The seasonal incidence of snakes at a locality in northern Belize. *Copeia* 1977:349–355.

Henderson, R.W., and M.A. Nickerson. 1976. Observations on the behavioral ecology of three species of *Imantodes* (Reptilia, Serpentes, Colubridae). *J. Herpetol.* 10:205–210.

Hillis, D.M., and R. de Sá. 1988. Phylogeny and taxonomy of the *Rana palmipes* group (Salientia: Ranidae). *Herpetol. Monogr.* 2:1–26.

Hillman, P.E. 1969. Habitat specificity in three sympatric species of *Ameiva* (Reptilia: Teiidae). *Ecology* 50:476–481.

Hirth, H.F. 1962. Food of *Basiliscus plumifrons* on a tropical strand. *Herpetologica* 18:276–277.

Hirth, H.F. 1963. Some aspects of the natural history of *Iguana iguana* on a tropical strand. *Ecology* 44:613–615.

Hirth, H.F. 1964. Observations on the fer-de-lance, *Bothops atrox,* in coastal Costa Rica. *Copeia* 1964:452–454.

Hoogmoed, M.S. 1973. Notes on the herpetofauna of Suriname. IV: The lizards and amphisbaenians of Suriname. The Hague: W. Junk Publishers.

Ibáñez, R. 1993. Female phonotaxis and call overlap in the Neotropical glassfrog *Centrolenella granulosa. Copeia* 1993:846–850.

Ibáñez, R.D., A.S. Rand, and C.A. Jaramillo. 1999. *The amphibians of Barro Colorado Nature Monument, Soberania National Park, and adjacent areas.* Santa Fe de Bogatá, Colombia: D´Vinni Editorial.

Iverson, J.B. 1982. Adaptations to herbivory in iguanine lizards. In *Iguanas of the world: Their behavior, ecology, and conservation,* edited by G.M. Burghardt and A.S. Rand. Park Ridge, N.J.: Noyes Publishing.

Iverson, J.B. 1992. *A revised checklist with distribution maps of the turtles of the world.* Richmond, Ind.: Privately published by author.

Janzen, D.H., editor. 1983. *Costa Rican natural history.* Chicago: University of Chicago Press.

Kitasako, J.T. 1967. Observations on the biology of *Dendrobates auratus* Schmidt and *Dendrobates pumilio* Girard. Master's thesis, University of Southern California.

Kluge, A.G. 1967. Higher taxonomic categories of gekkonid lizards and their evolution. *B. Am. Mus. Nat. Hist.* 135:1–59.

Kluge, A.G. 1981. The life history, social organization, and parental behavior of *Hyla rosenbergi* Boulenger, a nest-building gladiator frog. *Misc. Publ. Mus. Zool. Univ. Mich.,* no. 160.

Landy, M.J., D.A. Langebartel, E.O. Moll, and H.M. Smith. 1966. A collection of snakes from Volcan Tacana, Chiapas, Mexico. *J. Ohio Herpetol. Soc.* 5:93–101.

Layne, J.N., and T.M. Steiner. 1984. Sexual dimorphism in occurrence of keeled scales in the eastern indigo snakes *(Drymarchon corais couperi). Copeia* 1984:776–778.

Lee, J.C. 1996. *The amphibians and reptiles of the Yucatán Peninsula.* Ithaca, N.Y.: Cornell University Press.

Lee, J.C. 2000. *A field guide to the amphibians and reptiles of the Maya world.* Ithaca, N.Y.: Cornell University Press.

Leenders, T. 2001. *A guide to amphibians and reptiles of Costa Rica.* San José, Costa Rica: Distribuidores Zona Tropical.

Leenders, T., G. Beckers, and H. Strijbosch. 1996. *Micrurus mipartitus* (NCN): Polymorphism. *Herpetol. Rev.* 27:25.

Levy, C. 1991. *Endangered species: Crocodiles and alligators.* London: Quintet Publishing.

Levey, D.J., and F.G. Stiles. 1994. Birds: Ecology, behavior, and taxonomic affinities. In *La Selva: Ecology and natural history of a Neotropical rain forest,* edited by L.A. McDade, K.S. Bawa, H.A. Hespenheide, and G.S. Hartshorn. Chicago: University of Chicago Press.

Lieberman, S.S. 1986. Ecology of the leaf litter herpetofauna of a Neotropical rainforest: La Selva, Costa Rica. *Acta Zool. Mex.* 15: 1–72.

Limerick, S. 1976. Dietary differences of two sympatric Costa Rican frogs. Master's thesis, University of Southern California.

Liner, E.A. 1994. Scientific and common names for the amphibians and reptiles of Mexico in Spanish and English. *Soc. Study Amph. Rept. Herpetol. Circ.,* no. 23.

Lynch, J.D., and C.W. Myers. 1983. Frogs of the *fitzingeri* group of *Eleutherodactylus* in eastern Panama and Chocoan South America (Leptodactylidae). *B. Am. Mus. Nat. Hist.* 175:481–572.

Marquis, R.J., M.A. Donnelly, and C. Guyer. 1986. Aggregations of calling males of *Agalychnis calcarifer* Boulenger (Anura: Hylidae) in a Costa Rican lowland wet forest. *Biotropica* 18:173–175.

Martin, P.S. 1955. Herpetological records from the Gomez Farias region of southwestern Tamaulipas, Mexico. *Copeia* 1955:173–180.

Martinez, S., and L. Cerdas. 1986. Captive reproduction of the mussurana, *Clelia clelia* (Daudin), from Costa Rica. *Herpetol. Rev.* 17: 12–13.

McCoy, C.J. 1966. Additions to the herpetofauna of southern El Peten, Guatemala. *Herpetologica* 22:306–308.

McDade, L.A., and G.S. Hartshorn. 1994. La Selva Biological Station. In *La Selva: Ecology and natural history of a Neotropical rain forest,* edited by L.A. McDade, K.S. Bawa, H.A. Hespenheide, and G.S. Hartshorn. Chicago: University of Chicago Press.

McDiarmid, R.W. 1978. Evolution of parental care in frogs. In *The development of behavior: Comparative and evolutionary aspects,* edited by G.M. Burghardt and M. Bekoff. New York: Garland STPM Press.

McDiarmid, R.W., and R. Altig. 1999. *Tadpoles: The biology of anuran larvae.* Chicago: University of Chicago Press.

McVey, M.E., R.C. Zahary, D. Perry, and J. MacDougal. 1981. Territoriality and homing behavior in the poison dart frog *(Dendrobates pumilio). Copeia* 1981:1–8.

Miyamoto, M.M., and J.H. Cane. 1980a. Behavioral observations of non-calling males in Costa Rican *Hyla ebraccata. Biotropica* 12: 225–227.

Miyamoto, M.M., and J.H. Cane. 1980b. Notes on the reproductive behavior of a Costa Rican population of *Hyla ebraccata. Copeia* 1980:928–930.

Morales-Verdeja, S.A., and R. Vogt. 1997. *Kinosternon leucostomum* (pochitoque, chachagua). In *Historia natural de Los Tuxtlas,* edited by E. González Soriano, R. Dirzo, and R.C. Vogt. Mexico City: Universidad Nacional Autónoma de México.

Morris, M. 1991. Female choice of large males in the treefrog *Hyla ebraccata. J. Zool.* 223:371–378.

Muedeking, M.H., and W.R. Heyer. 1976. Description of eggs and reproductive patterns of *Leptodactylus pentadactylus* (Amphibia: Leptodactylidae). *Herpetologica* 32:137–139.

Myers, C.W. 1969a. Snakes of the genus *Coniophanes* in Panama. *Am. Mus. Novit.* 2374:1–28.

Myers, C.W. 1969b. The ecological geography of cloud forests in Panama. *Am. Mus. Novit.* 2396:1–52.

Myers, C.W. 1971. Central American lizards related to *Anolis pentaprion:* Two new species from the Cordillera de Talamanca. *Am. Mus. Novit.* 2471:1–40.

Myers, C.W. 1973. Anguid lizards of the genus *Diploglossus* in Panama, with the description of a new species. *Am. Mus. Novit.* 2523: 1–20.

Myers, C.W., and J.W. Daly. 1983. Dart-poison frogs. *Sci. Am.* 248: 120–133.

Myers, C.W., J.W. Daly, and B. Malkin. 1978. A dangerously toxic new frog *(Phyllobates)* used by Emberá Indians of western Colombia, with discussion of blow-gun fabrication and dart poisoning. *B. Am. Mus. Nat. Hist.* 161:307–366.

Neill, W.T. 1960. The caudal lure of various juvenile snakes. *Q. J. Fla. Acad. Sci.* 23:173–200.

Noble, G.K. 1918. The amphibians and reptiles collected by the American Museum expedition to Nicaragua in 1916. *B. Am. Mus. Nat. Hist.* 38:311–347.

Obst, F.J., K. Richter, and U. Jacob. 1988. *The completely illustrated atlas of reptiles and amphibians for the terrarium.* Neptune City, N.J.: T.F.H. Publications.

Oliver, J.A. 1947. The seasonal incidence of snakes. *Am. Mus. Novit.* 1363:1–14.

Perry, D.R. 1983. Access methods, observations, pollination biology, bee foraging behavior, and bee community structure within a Neotropical wet forest canopy. Ph.D. diss., University of California, Los Angeles.

Picado, C. 1976. *Serpientes venenosas de Costa Rica.* San José, Costa Rica: Universidad de Costa Rica.

Pombal, J.P., Jr., and M. Gordo. 1991. Duas novas especies de *Hyla* da Floresta Atlantica no Estado de Sao Paulo (Amphibia, Anura). *Mem. Inst. Butantan* 53:135–144.

Porras, L., J.R. McCranie, and L.D. Wilson. 1981. The systematics and distribution of the hognose viper *Bothrops nasuta* Bocourt (Serpentes: Viperidae). *Tulane Stud. Zool.* 22:85–107.

Pough, F.H. 1983. Amphibians and reptiles as low-energy systems. In *Behavioral energetics: The cost of survival in vertebrates,* edited by W.P. Aspey and S.I. Lustick. Columbus, Ohio: Ohio State University Press.

Pounds, J.A., and M.L. Crump. 1994. Amphibian declines and climate disturbance: The case of the golden toad and the harlequin frog. *Conserv. Biol.* 8:72–85.

Pyburn, W. T. 1970. Breeding behavior of the leaf-frogs *Phyllomedusa callidryas* and *Phyllomedusa dacnicolor* in Mexico. *Copeia* 1970: 209–218.

Ramirez, J., R. C. Vogt, and J. L. Villarreal-Benitez. 1997. *Rana vaillanti* (rana acuatica). In *Historia natural de Los Tuxtlas,* edited by E. González Soriano, R. Dirzo, and R. C. Vogt. Mexico City: Universidad Nacional Autónoma de México.

Rand, A. S. 1969. *Leptophis ahaetulla* eggs. *Copeia* 1969:402.

Rand, A. S. 1972. The temperatures of iguana nests and their relation to incubation optima and to nesting sites and season. *Herpetologica* 28:252–253.

Rand, A. S., B. A. Dugan, H. Monteza, and D. Vianda. 1990. The diet of a generalized folivore: *Iguana iguana* in Panama. *J. Herpetol.* 24: 211–214.

Rivero, J. A., and A. E. Esteves. 1969. Observations on the agonistic and breeding behavior of *Leptodactylus pentadactylus* and other amphibian species in Venezuela. *Breviora* 321:1–14.

Robakiewicz, P. E. 1992. Behavioral and physiological correlates of territoriality in a dart-poison frog *Dendrobates pumilio* Schmidt. Ph.D. diss., University of Connecticut, Storrs.

Roberts, W. E. 1994a. Evolution and ecology of arboreal egg-laying frogs. Ph.D. diss., University of California, Berkeley.

Roberts, W. E. 1994b. Explosive breeding, aggregations, and parachuting in a Neotropical frog, *Agalychnis saltator. J. Herpetol.* 28: 193–199.

Roberts, W. E. 1997. Behavioral observations of *Polychrus gutturosus,* a sister taxon of anoles. *Herpetol. Rev.* 26:184–185.

Rodda, G. H. 1992. The mating behavior of *Iguana iguana. Smithsonian Contrib. Zool.* 534:1–40.

Rodriguez, L. O., and W. E. Duellman. 1994. Guide to the frogs of the Iquitos region, Amazonian Peru. *Univ. Kans. Mus. Nat. Hist. Spec. Publ.,* no. 22.

Rosen, D. E. 1976. A vicariance model of Caribbean biogeography. *Syst. Zool.* 24:431–464.

Rosen, D. E. 1985. Geological hierarchies and biogeographic congruence in the Caribbean. *Ann. Missouri Bot. Gard.* 72:636–659.

Roze, J. A. 1984. New World coral snakes (Elapidae): A taxonomic and biological summary. *Mem. Inst. Butantan* 46:305–338.

Roze, J. A. 1996. *Coral snakes of the Americas: Biology, identification, and venoms.* Malabar, Fla.: Krieger Publishing.

Ruiz-Carranza, P. M., and J. D. Lynch. 1991. Ranas Centrolenidae de

Colombia. I. Propuesta de nueva clasificación genérica. *Lozania* 57:1–30.

Sanford, R.L., Jr., P. Paaby, J.C. Luvall, and E. Phillips. 1994. Climate, geomorphology, and aquatic systems. In *La Selva: Ecology and natural history of a Neotropical rain forest,* edited by L.A. McDade, K.S. Bawa, H.A. Hespenheide, and G.S. Hartshorn. Chicago: University of Chicago Press.

Savage, J.M. 1966. The origins and history of the Central American herpetofauna. *Copeia* 1966:719–766.

Savage J.M. 1968. The dendrobatid frogs of Central America. *Copeia* 1968:745–776.

Savage, J.M. 1982. The enigma of the Central American herpetofauna: Dispersals or vicariance? *Ann. Missouri Bot. Gard.* 69: 464–547.

Savage, J.M. 2002. *The amphibians and reptiles of Costa Rica: A herpetofauna between two continents, between two seas.* Chicago: University of Chicago Press.

Savage, J.M., and B.I. Crother. 1989. The status of *Pliocercus* and *Urotheca* (Serpentes: Colubridae), with a review of included species of coral snake mimics. *Zool. J. Linn. Soc.* 95:335–362.

Savage, J.M., and M.A. Donnelly. 1988. Variation and systematics in the colubrid snakes of the genus *Hydromorphus*. *Amphibia-Reptilia* 9:289–300.

Savage, J.M., and S.B. Emerson. 1970. Central American frogs allied to *Eleutherodactylus bransfordii* (Cope): A problem of polymorphism. *Copeia* 1970:623–644.

Savage, J.M., and K.R. Lips. 1993. A review of the status and biogeography of the lizard genera *Celestus* and *Diploglossus* (Squamata: Anguidae), with description of two new species from Costa Rica. *Rev. Biol. Trop.* 41:817–842.

Savage, J.M., and J.B. Slowinski. 1992. The colouration of the venomous coral snakes (family Elapidae) and their mimics (families Aniliidae and Colubridae). *Biol. J. Linn. Soc.* 45:235–254.

Savage, J.M., and J.B. Slowinski. 1996. Evolution of colouration, urotomy, and coral snake mimicry in the snake *Scaphiodontophis*. *Biol. J. Linn. Soc.* 57:129–194.

Savage, J.M., and J.L. Vial. 1974. The venomous coral snakes (genus *Micrurus*) of Costa Rica. *Rev. Biol. Trop.* 21:295–349.

Savage, J.M., and J. Villa. 1986. *SSAR contributions to herpetology.* No. 3, *Herpetofauna of Costa Rica.* Oxford, Ohio: Society for the Study of Amphibians and Reptiles.

Savitzky, A.H. 1981. Hinged teeth in snakes: An adaptation for swallowing hard-bodied prey. *Science* 212:346–349.

Scott, N.J., Jr. 1969. Zoogeographic analysis of snakes of Costa Rica. Ph.D. diss., University of Southern California.

Scott, N.J., Jr. 1976. The abundance and diversity of the herpetofauna of tropical forest litter. *Biotropica* 8:41–58.

Scott, N.J., Jr. 1983a. *Agalychnis callidryas* (rana calzonudo, gaudy leaf frog). In *Costa Rican natural history*, edited by D.H. Janzen. Chicago: University of Chicago Press.

Scott, N.J., Jr. 1983b. *Clelia clelia* (zopilota, musarana). In *Costa Rican natural history*, edited by D.H. Janzen. Chicago: University of Chicago Press.

Scott, N.J., Jr. 1983c. *Leptodactylus pentadactylus* (rana ternero, smoky frog). In *Costa Rican natural history*, edited by D.H. Janzen. Chicago: University of Chicago Press.

Scott, N.J., Jr. 1983d. *Rhadinaea decorata*. In *Costa Rican natural history*, edited by D.H. Janzen. Chicago: University of Chicago Press.

Scott, N.J., Jr., and S. Limerick. 1983. Reptiles and amphibians. In *Costa Rican natural history*, edited by D.H. Janzen. Chicago: University of Chicago Press.

Scott, N.J., Jr., J.M. Savage, and D.C. Robinson. 1983. Checklist of reptiles and amphibians. In *Costa Rican natural history*, edited by D.H. Janzen. Chicago: University of Chicago Press.

Seib, R.L. 1980. Human envenomation from the bite of an aglyphous false coral snake, *Pliocercus elapsoides* (Serpentes: Colubridae). *Toxicon* 18:399–401.

Seib, R.L. 1984. Prey use in three sympatric Neotropical racers. *J. Herpetol.* 18:412–420.

Seib, R.L. 1985a. Europhagy in a tropical snake, *Coniophanes fissidens*. *Biotropica* 17:57–64.

Seib, R.L. 1985b. Feeding ecology and organization of Neotropical snake faunas. Ph.D. diss., University of California, Berkeley.

Seifert, R.P. 1983. *Bothrops schlegelii*. In *Costa Rican natural history*, edited by D.H. Janzen. Chicago: University of Chicago Press.

Sexton, O.J., and H. Heatwole. 1965. Life history notes on some Panamanian snakes. *Caribb. J. Sci.* 5:39–43.

Silverstone, P.A. 1975. A revision of the poison-arrow frogs of the genus *Dendrobates* Wagler. *Nat. Hist. Mus. Los Angeles Co. Sci. Bull.* 21: 1–55.

Silverstone, P.A. 1976. A revision of the poison-arrow frogs of the genus *Phyllobates* Bibron in Sagra (Dendrobatidae). *Nat. Hist. Mus. Los Angeles Co. Sci. Bull.* 27:1–53.

Slevin, J. R. 1942. Notes on a collection of reptiles and amphibians from Guatemala. II. Lizards. *Proc. Calif. Acad. Sci., Ser. 4* 23: 453–462.

Slowinski, J. B., and J. M. Savage. 1995. Urotomy in *Scaphiodontophis:* Evidence for the multiple tail break hypothesis. *Herpetologica* 51:338–341.

Slowinski, J. B., B. I. Crother, and J. E. Fauth. 1987. Diel differences in leaf-litter abundances of several species of reptiles and amphibians in an abandoned cacao grove in Costa Rica. *Rev. Biol. Trop.* 35:349–350.

Smith, H. M. 1943. Summary of collections of snakes and crocodilians made in Mexico under the Walter Rathborne Bacon Traveling Scholarship. *Proc. U. S. Natl. Mus.* 93:393–504.

Smith, H. M. 1968. A new pentaprionid anole (Reptilia: Lacertilia) from Pacific slopes of Mexico. *Trans. Kans. Acad. Sci.* 71:195–200.

Smith, H. M., and C. Grant. 1958. New and noteworthy snakes from Panama. *Herpetologica* 14:207–215.

Sollins, P., F. Sancho M., R. Mata Ch., and R. L. Sanford, Jr. 1994. Soils and soil process research. In *La Selva: Ecology and natural history of a Neotropical rain forest,* edited by L. A. McDade, K. S. Bawa, H. A. Hespenheide, and G. S. Hartshorn. Chicago: University of Chicago Press.

Solórzano, A., and L. Cerdas. 1987. *Drymobius margaritiferus* (speckled racer): Reproduction. *Herpetol. Rev.* 18:75–76.

Solórzano, A., and L. Cerdas. 1989. Reproductive biology and distribution of the terciopelo, *Bothrops asper* Garman (Serpentes: Viperidae) in Costa Rica. *Herpetologica* 45:444–450.

Strieby, A. M. 1998. A multispecific frog chorus: Calls, calling sites, and acoustic interference in a complex forest environment. Master's thesis, California State University, Northridge.

Stuart, L. C. 1935. A contribution to a knowledge of the herpetofauna of a portion of the savanna region of central Petén, Guatemala. *Misc. Publ. Mus. Zool. Univ. Mich.,* no. 29.

Stuart, L. C. 1948. The amphibians and reptiles of Alta Verapaz, Guatemala. *Misc. Publ. Mus. Zool. Univ. Mich.,* no. 69.

Stuart, L. C. 1966. The environment of the Central American cold-blooded vertebrate fauna. *Copeia* 1966:684–699.

Swanson, P. L. 1945. Herpetological notes from Panama. *Copeia* 1945:210–216.

Swanson, P. L. 1950. The iguana *Iguana iguana iguana* (L.). *Herpetologica* 6:187–193.

Talbot, J.J. 1977. Habitat selection in two tropical anoline lizards. *Herpetologica* 33:114–123.

Talbot, J.J. 1979. Time budget, niche overlap, inter- and intraspecific aggression in *Anolis humilis* and *Anolis limifrons* from Costa Rica. *Copeia* 1979:472–481.

Taylor, E.H. 1951. A brief review of the snakes of Costa Rica. *Univ. Kans. Sci. Bull.* 34:3–188.

Taylor, E.H. 1952. A review of frogs and toads of Costa Rica. *Univ. Kans. Sci. Bull.* 35:577–942.

Taylor, E.H. 1955. Additions to the known herpetological fauna of Costa Rica with comments on other species, no. 2. *Univ. Kans. Sci. Bull.* 37:499–575.

Taylor, E.H. 1958. Additions to the known herpetological fauna of Costa Rica with comments on other species, no. 3. *Univ. Kans. Sci. Bull.* 39:3–40.

Terborgh, J. 1990. An overview of research at Cocha Cashu Biological Station. In *Four Neotropical rainforests,* edited by A.H. Gentry. New Haven, Conn.: Yale University Press.

Test, F.H., O.J. Sexton, and H. Heatwole. 1966. Reptiles of Rancho Grande and vicinity, Estato Aragua, Venezuela. *Misc. Publ. Mus. Zool. Univ. Mich.,* no. 128.

Thorbjarnarson, J.B. 1993a. Diet of the spectacled caiman *(Caiman crocodylus)* in the central Venezuelan llanos. *Herpetologica* 49: 108–117.

Thorbjarnarson, J.B. 1993b. Fishing behavior of spectacled caiman in the Venezuelan llanos. *Copeia* 1993: 1166–1171.

Thorbjarnarson, J.B. 1994. Reproductive ecology of the spectacled caiman *(Caiman crocodylus)* in the Venezuelan llanos. *Copeia* 1994:907–919.

Thorbjarnarsen, J.B. 1995. Dry season diel activity patterns of spectacled caiman *(Caiman crocodylus)* in the Venezuelan llanos. *Amphibia-Reptilia* 16:415–421.

Toft, C.A. 1981. Feeding ecology of Panamanian litter anurans: Patterns in diet and foraging mode. *J. Herpetol.* 15:139–144.

Townsend, D.S. 1996. Patterns of parental care in frogs of the genus *Eleutherodactylus.* In *Contributions to herpetology.* Vol. 12, *Contributions to West Indian herpetology: A tribute to Albert Schwartz,* edited by R. Powell and R.W. Henderson. Ithaca, N.Y.: Society for the Study of Amphibians and Reptiles.

Troyer, K. 1982. Transfer of fermentative microbes between generations in a herbivorous lizard. *Science* 216:540–542.

van Berkum, F. H. 1986. Evolutionary patterns of the thermal sensitivity of sprint speed in *Anolis* lizards. *Evolution* 40:594–604.

van Berkum, F. H. 1988. Latitudinal patterns of the thermal sensitivity of sprint speed in lizards. *Am. Nat.* 132:327–343.

Vandermeer, J. H., J. Stout, and S. Risch. 1979. Seed dispersal of a common Costa Rican rainforest palm *(Welfia georgii)*. *Trop. Ecol.* 20:17–26.

Vanzolini, P. E. 1983. Guiano-Brazilian *Polychrus:* Distribution and speciation (Sauria: Iguanidae). In *Advances in herpetology and evolutionary biology: Essays in honor of Ernest E. Williams,* edited by A. G. J. Rhodin and K. Miyata. Cambridge, Mass.: Museum of Comparative Zoology.

Villa, J. 1973. A snake in the diet of a kinosternid turtle. *J. Herpetol.* 7:380–381.

Villarreal-Benitez, J. L. 1997. Historia natural del genero *Anolis.* In *Historia natural de Los Tuxtlas,* edited by E. González Soriano, R. Dirzo, and R. C. Vogt. Mexico City: Universidad Nacional Autónoma de México.

Vinton, K. W. 1951. Observations on the life history of *Leptodactylus pentadactylus. Herpetologica* 7:73–75.

Vitt, L. J., and S. de La Torre. 1996. Guia para la investigacion de las lagartijas de Cuyabeno. *Mus. Zool. Pontifica Univ. Catolica Ecuador Monogr.,* no. 1.

Vitt, L. J., and T. E. Lacher. 1981. Behavior, habitat, diet, and reproduction of the iguanid lizard *Polychrus marmoratus* in the caatinga of northeastern Brazil. *Herpetologica* 37:53–63.

Vitt, L. J., and P. A. Zani. 1996. Ecology of the lizard *Ameiva festiva* (Teiidae) in southeastern Nicaragua. *J. Herpetol.* 30:110–117.

Vitt, L. J., and P. A. Zani. 1997. Ecology of the nocturnal lizard *Thecadactylus rapicauda* (Sauria: Gekkonidae) in the Amazon region. *Herpetologica* 53:165–179.

Vitt, L. J., P. A. Zani, and R. D. Durtsche. 1995. Ecology of the lizard *Norops oxylophus* (Polychrotidae) in lowland forest of southeastern Nicaragua. *Can. J. Zool.* 73:1918–1927.

Vogt, R. C., and O. F. Flores-Villela. 1992. Effects of incubation temperature on sex determination in a community of Neotropical freshwater turtles in southern Mexico. *Herpetologica* 48:265–270.

Vogt, R. C., and S. Guzman. 1988. Food partitioning in a Neotropical freshwater turtle community. *Copeia* 1988:37–47.

Wake, M. H. 1983. *Gymnopis multiplicata, Dermophis mexicanus,* and *Dermophis parviceps* (Soldas, sulda con sulda, dos cabezas, caecilians). In *Costa Rican natural history,* edited by D. H. Janzen. Chicago: University of Chicago Press.

Warkentin, K. M. 1995. Adaptive plasticity in hatching age: A response to predation risk trade-off. *Proc. Natl. Acad. Sci. U.S.A.* 92: 3507–3510.

Watling, J. I., and M. A. Donnelly. 2002. Seasonal patterns of reproduction and abundance of leaf litter frogs in a Central American rainforest. *J. Zool.* 258:269–276.

Webb, R. G. 1958. The status of the Mexican lizards of the genus *Mabuya. Univ. Kans. Sci. Bull.* 38:1303–1313.

Wells K. D., and J. J. Schwartz. 1984. Interspecific acoustic interactions of the Neotropical frog, *Hyla ebraccata. Behav. Ecol. Sociobiol.* 14:211–224.

Werman, S. D. 1992. Phylogenetic relationships of Central and South American pit vipers of the genus *Bothrops* (sensu lato): Cladistic analyses of biochemical and anatomical characters. In *Biology of the pitvipers,* edited by J. A. Campbell and E. D. Brodie, Jr. Tyler, Tex.: Selva Press.

Weyer, D. 1990. *Snakes of Belize.* Belize City: Angelus Press.

Weygoldt, P. 1980. Complex brood care and reproductive behavior in captive poison-arrow frogs, *Dendrobates pumilio* O. Schmidt. *Behav. Ecol. Sociobiol.* 7:329–332.

Williams, E. E. 1983. Ecomorphs, faunas, island size, and diverse end points in island radiations of *Anolis.* In *Lizard ecology: Studies of a model organism,* edited by R. B. Huey, E. R. Pianka, and T. W. Schoener. Cambridge, Mass.: Harvard University Press.

Williams, E. E. 1984. New or problematic *Anolis* from Colombia. III. Two new semiaquatic anoles from Antioquia and Choco, Colombia. *Breviora* 478:1–22.

Wilson, L. D. 1987. A resumé of the colubrid snakes of the genus *Tantilla* of South America. *Milwaukee Publ. Mus. Contrib. Biol. Geol.* 68:1–35.

Wilson, L. D., and J. R. Meyer. 1985. *The snakes of Honduras.* Milwaukee, Wis.: Milwaukee Public Museum.

Winter, Y. 1987. A comparative study of the locomotor rhythms of 11 species of frogs from a Costa Rican rainforest. Master's thesis, University of Minnesota.

Wollerman, L. 1995. Acoustic communication and acoustic interference in a Neotropical frog, *Hyla ebraccata.* Ph.D. diss., University of North Carolina.

Wollerman, L. 1998. Stabilizing and directional preferences of female

Hyla ebraccata for calls in static properties. *Anim. Behav.* 55: 1619–1630.

Zaher, H. 1994. Les Tropidopheoidea (Serpentes: Alethinophidia) sont-ils réellement monophylétiques? Arguments en faveur de leur polphylétisme. *C. R. Acad. Sci. Paris* 317:471–487.

Zamudio, K. R., and H. W. Greene. 1997. Phylogeography of the bushmaster (*Lachesis muta:* Viperidae): Implications for Neotropical biogeography, systematics, and conservation. *Biol. J. Linn. Soc.* 62:421–442.

Zimmerman, B. L., and R. O. Bierregaard. 1986. Relevance of the equilibrium theory of island biogeography and species-area relations to conservation with a case from Amazonia. *J. Biogeogr.* 13: 133–143.

Zug, G. R., S. B. Hedges, and S. Sunkel. 1979. Variation in reproductive parameters of three Neotropical snakes, *Coniophanes fissidens, Dipsas catesbyi,* and *Imantodes cenchoa. Smithsonian Contrib. Zool.* 300:1–20.

ART CREDITS

Color Plates

K. A. BAKKEGARD 65, 94, 102, 104, 110, 112, 150, 154, 177, 179, 192

EMMETT L. BLANKENSHIP 1, 4, 11, 12, 14, 30, 48, 52, 64, 68, 80, 87, 100, 132, 138, 139, 147, 153, 163, 167, 182, 184

S. M. BOBACK 13, 15–19, 22, 26, 66, 88, 111, 119, 148, 160, 162, 165, 176, 186, 189, 199, 207

P. A. BUTTENHOFF 5, 6, 25, 33, 35, 49, 50, 53, 63, 69, 90, 95, 96, 103, 117, 122, 130, 131, 136, 140, 149, 158, 191

R. COLEMAN 129

M. A. DONNELLY 2, 3, 10, 23, 24, 28, 29, 32, 34, 40, 56, 59, 67, 70, 71, 73, 75, 76, 78, 79, 83, 91–93, 106, 109, 126, 128, 133, 134, 144, 152, 155, 156, 173–175, 181, 183, 188, 190, 200, 203, 205, 206

C. D'ORGEIX 41, 42, 74, 98, 99, 105, 180

C. GUYER 7, 21, 31, 37–39, 43–47, 54, 55, 57, 60–62, 72, 81, 82, 84–86, 101, 107, 113–116, 120, 121, 123–125, 142, 143, 151, 166, 169, 171, 172, 185, 187, 193, 197, 201

S. M. HERMANN 8, 9, 97, 127

C. J. LEARY 20, 137, 146

M. LUNDBERG 77

R. W. MCDIARMID 198, 202

C. W. MYERS 195

K. E. NICHOLSON 108

J. R. PARMELEE 145, 157, 159, 168, 170, 178, 204

R. N. REED 89, 135

B. ROGELL 164

G. G. SORRELL 196

D. L. WAGNER 27, 36, 51, 58

M. I. WILLIAMS 118, 141, 161, 194

Line Illustrations

BILL NELSON Figure 1

CRISTINA UGARTE Figures 2–20

INDEX

Page numbers in **bold** refer to main discussion of the topic.

acarines, 42, 77, 79, 82, 88, 94, 95, 97, 136, 138
Adelia triloba, 10
Agalychnis
  calcarifer, 15, 46, **47–48**
  callidryas, 46, **49–50,** 244
  saltator, 46, 49, **50–51**
  spurrelli, 50, **244**
agouti, 161
Agouti paca, 161
ajillo, 9, 10
Allen's Coralsnake, 190, 195, 213, 220, **227–228**
Alligatoridae, **234–235**
almendro, 9
Alvarado's Salamander, 11, 16, 31, **241–242**
Amastridium veliferum, 176, **179–180,** 189, 191
Amblyomma, 115, 116
Ameiva, 170
  festiva, 15, **154–155,** 203, 207, 209, 210, 247
  quadrilineata, 16, 155, **247**
American Crocodile, 235, 236–237
Amphibia, 22–102, 241–245
Angiospermae
  Adelia triloba, 10
  Asplundia uncinata, 56
  Asterogyne, 9
  Astrocaryum alatum, 10
  Bactris, 9

Bactris gasipaes, 134, 135, 137
Bactris longiseta, 10
Carapa nicaraguensis, 10
Chione costaricensis, 10
Cordia alliodora, 134, 135, 137
Dendropanax arboreus, 9
Dipteryx panamensis, 9
Euterpe macrospadix, 9
Ficus insipida, 10
Geonoma, 9
Grias cauliflora, 10
Heliconia, 123, 166
Inga ruiziana, 10
Iriartea deltoidea, 9
Luehea seemannii, 10
Melastomataceae, 9
Myrica splendens, 10
Otoba novogranatensis, 10
Pachira aquatica, 10
Panicum grande, 10
Piperaceae, 9
Pithecellobium elegans, 9
Pithecellobium longifolium, 10
Protium pittieri, 9
Psychotria chagrensis, 10
Pterocarpus officinalis, 10
Rubiaceae, 9
Socrotea exorrhiza, 9
Spathiphyllum freidrichsthalli, 10
Unonopsis pittieri, 9
Warscewiczia coccinea, 9
Welfia georgii, 9

Anguidae, 120, 156–159
Annulated Tree Boa, 161–162
anoles, 128–129, 141, 193, 207
    Carpenter's Anole, **133–134**
    Dry Forest Anole, 11
    Green Tree Anole, **131–132**
    Ground Anole, **135–136**
    Lemur Anole, **136–137,** 246
    Lichen Anole, **140–141**
    Puerto Rican Crested Anole, 137,
        246
    Pug-nosed Anole, **132–133**
    Slender Anole, **137–138**
    Stream Anole, **139**
Anomalepididae, 247
Antbird, 161
ants, 42, 124
anturto blanco, 10
Anura, 23, 25, 28, 33–102, 242–245
apterygotes, 147
arachnids, 132
    nonspider, 42, 77, 82, 88, 89, 94
araneids, 42, 77, 79, 82, 84, 88, 89, 94,
        97, 136, 137, 138
Armadillo, Nine-banded, 182
Arremon aurantiirostris, 161
arthropods, 31, 32, 41, 42, 48, 49, 51,
        52, 54, 55, 56, 57, 58, 60, 61, 62,
        63, 67, 68, 69, 70, 7, 76, 77, 82, 93,
        84, 86, 87, 88, 89, 90, 101, 134,
        135, 139, 140, 146, 147, 148, 149,
        153, 155, 157, 158
Asplundia uncinata, 56
Asterogyne, 9
Astrocaryum alatum, 10
Aves
    Arremon aurantiirostris, 161
    Glyphorhynchus spirurus, 170
Bactris
    gasipaes, 134, 135, 137
    longiseta, 10
bananas, 11
Basiliscus, 122
    plumifrons, **122–123,** 127
    vittatus, 18, **124–125**
basilisks
    Green Basilisk, **122–123,** 127
    Striped Basilisk, 18, **124–125**

bats, 35, 123, 161, 211
beetles, 126, 149
Bicolored Snaileater, **186**
Bird-eating Snake, 188–189, **210–211**
birds
    Antbird, 161
    Blue-gray Tanager, 161
    hummingbirds, 162, 166
    Orange-billed Sparrow, 161
    poultry, 161
    Wedge-billed Woodcreeper, 170
    White-collared Swift, 211
Black Wood Turtle, 114, **115–116**
Black-tailed Cribo, **186–187**
blattids, 138, 155
Blindsnake, Pink-headed, **247**
Blue-gray Tanager, 161
boas
    Annulated Tree Boa, **161–162**
    Boa Constrictor, **160–161**
    Rainbow Boa, **247**
Boa constrictor, 159, **160–161**
Boidae, 121, **159–161, 247**
bola de oro, 10, 247
Bolitoglossa, 31
    alvaradoi, 11, 31, **241–242**
    colonnea, 29, **30–31,** 242
    striatula, 31, **242**
Bothriechis schlegelii, 164, **165–166**
Bothrops, 164
    asper, **166–167,** 168
Boulenger's Snouted Treefrog, **57–58,**
        59
Bransford's Litterfrog, **76–77,** 84, 155,
        185, 188, 200, 206, 212
Brilliant Forest Frog, **101–102,** 170
Broad-headed Rainfrog, **84,** 224, 226,
        242
bromeliads, 28, 31, 35, 62, 82, 97, 193,
        199, 200
Bronze-backed Climbing Skink,
        152–153, 157
Brown Blunt-headed Vinesnake,
        193–194, 195
Brown Debris Snake, **183–184**
Brown Forest Racer, 181, 184, 185,
        202–203
Brown Vinesnake, **206–207**

Brown Wood Turtle, **114–115**
Bufo
  coniferus, **40–41**, 44
  haematiticus, 39, **41–42**
  marinus, 40, **42–43**, 44
  melanochlorus, 40, 41, **44–45**
Bufonidae, 36, **39–44**
Bushmaster, 17, 18, 164, 166, **168–169**
butterflies, 15
cacao, 11, 67, 96, 133, 134, 135, 137,
    146, 148, 153, 184, 218, 223
caecilians, 6, 22, 23, **24–26, 241**
  La Loma Caecilian, 26, **241**
  Purple Caecilian, **25–26**, 241
Caeciliidae, **25–26, 241**
cafecillo, 10
Caiman, Spectacled, 17, **234–235**, 236
Caiman crocodylus, 17, **234–235**, 236
Calico Snake, 190, 195, **209–210**, 213,
    220, 224, 228, 229, 230
Carapa nicaraguensis, 10
Carpenter's Anole, **133–134**
Casque-headed Lizard, **125–126**
caterpillars, 126
cat-eyed snakes
  Northern Cat-eyed Snake, 50,
      197–198, 248
  Southern Cat-eyed Snake, **196–197**
Caudata, 23, **27–32, 241–242**
Celestus, Rainforest, 152, 153,
    156–157
Celestus
  hylaius, 152, 153, **156–157**
  cyanochloris, 157
centipedes, 77, 80, 85, 94, 136, 151,
    154, 218, 219
Central American Coralsnake, 190,
    213, 220, 228, **229–231,** 248
Central American Coralsnake Mimic,
    190–191, 195, 205, 210, 213, 220,
    228, 230
Central American Dwarf Boa, **163**
Central American Whiptail, 15,
    154–155, 203, 207, 209, 210, 247
Centrolenella, 67, 68, 69, 72
  ilex, 72, **245**
  prosoblepon, **65–66**
Centrolenidae, 38, **64–72, 245**

Chameleon, Neotropical, **141–142**
Chelydra serpentina, **109–110**
Chelydridae, 107, **108–110**
chilamate, 10
chilopods, 95
Chione costaricensis, 10
Chironius grandisquamis, 176,
    180–181, 182, 217
chocolate, 11
clavillo, 10
Clay-colored Rainfrog, **78–79,** 81, 88
Clelia clelia, 175, 176, **181–182,** 204,
    220
Cloudy Slugeater, 204, **216**
Coati, White-nosed, 17, 161
Cochranella
  albomaculata, 65, **67**, 70, 71
  granulosa, 65, **68**
  spinosa, 65, 66, **69**
Coendou rothschildi, 161
Coffeesnake, Red, **204–205**
cola de gallo, 56
coleopterans, 26, 42, 58, 66, 77, 79, 82,
    84, 88, 92, 94, 95, 97, 100, 132,
    135, 137, 138, 139, 140
collembolans, 95
Colostethus, 17
Colubridae, 120, **171–226,** 248–249
Common Mexican Treefrog, **59–60,**
    181, 199
Common Snapping Turtle, **109–110**
Common Tink Frog, **81–82,** 212, 214
Coniophanes fissidens, 178, **183–184**
consuelo de mujer, 9
copal, 9
Corallus, 160
  annulatus, 159, **161–162,** 247
Coral-mimic Galliwasp, 157,
    158–159
coralsnakes
  Allen's Coralsnake, 190, 195, 213,
      220, **227–228**
  Central American Coralsnake, 190,
      213, 220, 228, **229–231,** 248
  Many-banded Coralsnake, 190,
      213, 220, 228, **229–231**
Coral-spotted Rainfrog, **75–76**
Cordia alliodora, 134, 135, 137

Corytophanes, 122
  cristatus, 122, **125–126**
Corytophanidae, 120, **122–126**
Costa Rica feather palm, 9
Costa Rican fruta de pava, 10
coyolillo, 10
crabs, 159
Crested Anole, Puerto Rican, 137, **246**
Cribo, Black-tailed, **186–187**
Crocodile, American, 16, **236–237**
crocodilians, 6, 18, **233–237**
Crocodylia, 105, **233–237**
Crocodylidae, **236–237**
Crocodylus acutus, 16, **236–237**
crowned snakes
  Orange-bellied Crowned Snake,
    218–219, 221, 225
  Reticulated Crowned Snake,
    217–218, 219, 221, 225
  Tricolored Crowned Snake, 190,
    195, 210, 213, **219–220,** 228, 230
Ctenonotus cristatellus, 137, **246**
Ctenosaura similis, 161
dart-poison frogs, **94–99**
  Striped Dart-poison Frog, **97–99,**
    242, 243
Dasypus novemcinctus, 182
deer, 161, 233
Degenhardt's Scorpion-eater, 16, 198,
  248–249
Dendrobates
  auratus, 16, **95–96**
  pumilio, 12, 95, **96–97**
  typographicus, 12
Dendrobatidae, 36, **94–99**
Dendropanax arboreus, 9
Dendrophidion
  percarinatum, 178, 179, **184,** 185,
    203
  vinitor, 178, 184, **185,** 203
Dermophis parviceps, 26, **241**
dermopterans, 77, 84
Diamondback Racer, 16, 198, **248**
Didelphis, 167
Diploglossus
  bilobatus, 156, **157–158**
  monotropis, 156, 157, **158–159**
Dipsas bicolor, 171, **186**

dipterans, 77, 78, 79, 80, 82, 84, 89, 94,
  97, 133, 138, 139, 140, 147, 153
Dipteryx panamensis, 9
Drab Treefrog, **63–64**
Dry Forest Anole, 11
Drymarchon corais, **186–187**
Drymobius
  margaritiferus, 177, **187–188**
  melanotropis, 174, **188–189**
  rhombifer, 16, 198, **248**
Dwarf Boa, Central American, 163
dwarf geckos
  Spotted Dwarf Gecko, 146,
    147–148
  Yellow-tailed Dwarf Gecko, 145,
    146, 147, **148**
dwarf snakes
  Faded Dwarf Snake, 218, 219,
    221–222, 223, 225
  Viquez's Dwarf Snake, 212, 222,
    249
earthsnakes
  Hoffmann's Earthsnake, 176, 179,
    189, **191**
  Ruthven's Earthsnake, 16
earthworms, 24, 26, 191, 204, 205
Ebony Keelback, **180–181,** 182, 217
Elapidae, 120, **226–231**
Eleutherodactylus, 38, 73, 101, 170, 201
  altae, 74, **75–76**
  biporcatus, 84
  bransfordii, 74, **76–77,** 84, 155, 185,
    188, 200, 206, 212
  bufoniformis, 84, **242**
  caryophyllaceus, 74, 76, **77–78**
  cerasinus, 74, **78–79,** 81, 88
  crassidigitus, 74, **79–80,** 83, 86, 89
  cruentus, 74, 76, 79, **80–81,** 88
  diastema, 74, **81–82,** 212, 214
  fitzingeri, 73, 80, **82–83,** 86, 89,
    188, 198
  gaigeae, 98, **242–243**
  gollmeri, 86, **243**
  megacephalus, 74, **84,** 224, 226, 242
  mimus, 74, **85–86,** 205, 224, 243
  noblei, 74, 85, **86**
  ranoides, 74, **87,** 90
  ridens, 74, 79, 81, 82, **88**

rugulosus, 87
talamancae, 74, 80, 83, 86, **89**
Emerald Glassfrog, **65–66**
Emydidae, 107, **113–116, 246**
Enulius sclateri, 176, 179, 180, **189,** 191
Epicrates, 162
cenchria, **247**
epiphytic plants, 8, 28, 31, 35, 82, 162
Erythrolamprus mimus, 173, **190–191,**
195, 205, 210, 213, 220, 228, 230
Euterpe macrospadix, 9
Eyelash Viper, **165–166**
Faded Dwarf Snake, 218, 219,
221–222, 223, 225
False Fer-de-Lance, 166, **225–226**
False Tree Coral, 230–231, **248**
Felis pardalis, 161
ferns, 35, 115
Ficus insipida, 10
Fire-bellied Snake, **201–202**
fish, 17, 22, 27, 28, 45, 53, 56, 106, 107,
192, 221, 235, 236
Fitzinger's Rainfrog, 80, **82–83,** 86, 89,
188, 198
Fleischmann's Glassfrog, 67, **70–71**
foamfrogs
Fringe-toed Foamfrog, **90–91**
Turbo White-lipped Foamfrog, 91,
243–244
Forest Frog, Brilliant, **101–102,** 170
forest racers
Brown Forest Racer, **184,** 185
Lowland Forest Racer, 184, **185**
formicids, 26, 77, 82, 84, 88, 89, 91, 95,
97, 98, 133, 135, 139, 140, 151
Four-lined Whiptail, 16, 155, **247**
Fringe-toed Foamfrog, **90–91**
frogs
Brilliant Forest Frog, **101–102,** 170
Common Tink Frog, **81–82,** 212,
214
Leopard Frog, 101
Smoky Jungle Frog, **91–92**
Taylor's Frog, 101, **245**
Tink Frog, **81–82,** 212, 214
Vaillant's Frog, 91, **99–101,** 197,
199, 203
See also dart-poison frogs; glass-

frogs; leaf frogs; Litterfrog,
Bransford's; poison frogs; rainfrogs; Sheepfrog, Reticulated;
treefrogs
fruta de pava, 10
fungus, 70
fungus-flies, 70
Gaige's Rainfrog, 98, **242–243**
galliwasps
Coral-mimic Galliwasp, 157,
158–159
Talamancan Galliwasp, **157–158**
Gastrophryne pictiventris, **93–94,** 181,
188
gastropods, 186
gavilán, 9
geckos
Litter Gecko, 145, **146–147,** 148,
222
Spotted Dwarf Gecko, 146,
147–148
Turnip-tailed Gecko, **149–150**
Yellow-headed Gecko, 16, **145–146,**
148
Yellow-tailed Dwarf Gecko, 145,
146, 147, **148**
Gekkonidae, 120, **143–150**
Geonoma, 9
Geophis, 191
hoffmanni, 176, 179, 189, **191**
ruthveni, 16
Ghost Glassfrog, 72, **245**
glassfrogs
Emerald Glassfrog, **65–66**
Fleischmann's Glassfrog, 67, **70–71**
Ghost Glassfrog, 72, **245**
Granular Glassfrog, **68**
Powdered Glassfrog, 67, **71–72**
Reticulated Glassfrog, **72**
Spined Glassfrog, 66, **69**
Yellow-flecked Glassfrog, **67,** 70, 71
Glyphorhynchus spirurus, 170
Golden-groined Rainfrog, 76, 79,
80–81, 88
Gonatodes albogularis, 16, 144,
145–146, 148
Granular Glassfrog, **68**
grasshoppers, 80, 142

Gray-eyed Leaf Frog, 15, **47–48**
Green and Black Poison Frog, **95–96**
Green Basilisk, **122–123,** 127
Green Climbing Toad, **40–41,** 44
Green Iguana, **126–128,** 161
Green Parrotsnake, **198–199,** 200
Green Tree Anole, **131–132**
Green Vinesnake, 199, **208–209**
Grias cauliflora, 10
Ground Anole, **135–136**
guácimo colorado, 10
Gymnophiona, **24–26, 241**
Gymnopis multiplicata, **25–26,** 241
Halloween Snake, 171, 210, **223–224,**
    229
Harlequin Treefrog, **52–53,** 54, 55, 185
Heliconia, 123, 166
Helminthophis frontalis, **247**
hemipterans, 77, 81, 85, 94, 136, 138,
    147, 154
Heteromys, 196
    desmarestianus, 170
Highland Fringe-limbed Treefrog, 61,
    244
Hoffman's Earthsnake, 176, 179, 189,
    191
Hog-nosed Viper, **169–170**
homopterans, 77, 84, 88, 97, 134, 136,
    153
huevos de burro, 10
hummingbirds, 162, 166
Hyalinobatrachium
    fleischmanni, 65, 67, **70–71**
    pulveratum, 65, 67, **71–72**
    valerioi, 65, **72**
Hydromorphus concolor, 177, **192**
Hyla
    ebraccata, 46, **52–53,** 54, 55, 185
    loquax, 46, **53–54**
    miliaria, 61, **244**
    phlebodes, 46, 52, 53, **54–55,** 58
    rufitela, 46, **55–56**
Hylidae, 38, **45–64, 244**
hymenopterans, 42, 81, 82, 84, 88, 94,
    153
iguanas, 18
Iguana iguana, 123, **126–128,** 161
Iguanidae, 7, 120, **126–128**

Imantodes, 207
    cenchoa, 178, **193–194,** 195
    inornatus, 178, **194–195,** 193
Inga ruiziana, 10
insects
    ants, 42, 124
    apterygotes, 147
    beetles, 126, 149
    blattids, 138, 155
    butterflies, 15
    caterpillars, 126
    coleopterans, 26, 42, 58, 66, 77, 79,
        82, 84, 88, 92, 94, 95, 97, 100, 132,
        135, 137, 138, 139, 140
    collembolans, 95
    dipterans, 77, 78, 79, 80, 82, 84, 89,
        94, 97, 133, 138, 139, 140, 147,
        153
    formicids, 26, 77, 82, 84, 88, 89, 91,
        95, 97, 98, 133, 135, 139, 140, 151
    fungus-flies, 70
    grasshoppers, 80, 142
    hemipterans, 77, 81, 85, 94, 136,
        138, 147, 154
    homopterans, 77, 84, 88, 97, 134,
        135–136, 153
    hymenopterans, 42, 81, 82, 84, 88,
        94, 154
    isopterans, 85, 94, 151
    lepidopterans, 78, 85, 95, 137, 139
    moths, 15, 149
    orthopterans, 66, 77, 78, 80, 82, 84,
        89, 94, 97, 137, 138, 147, 153, 155
    ponerine ants, 94
    tettegoniids, 126
invertebrates
    acarines, 42, 77, 79, 82, 88, 94, 95,
        97, 136, 138
    arachnids, 132
    araneids, 42, 77, 79, 82, 84, 88, 89,
        94, 97, 136, 138
    arthropods, 31, 32, 41, 42, 48, 49,
        51, 52, 54, 55, 56, 57, 58, 60, 61,
        62, 63, 67, 68, 69, 70, 7, 76, 77, 82,
        93, 84, 86, 87, 88, 89, 90, 101, 134,
        135, 139, 140, 146, 147, 148, 149,
        153, 155, 157, 158

centipedes, 77, 80, 85, 94, 136, 151, 154, 218, 219
chilopods, 95
earthworms, 24, 26, 191, 204, 205
gastropods, 186
isopods, 77, 78, 80, 82, 84, 85, 88, 89, 97, 136, 138, 147, 154, 155
land crabs, 159
leeches, 205
nonspider arachnids, 42, 77, 82, 88, 89, 94
oligochaetes, 26, 94
slugs, 204, 205, 214, 216
ticks, 115, 116, 135
Iriartea deltoidea, 9
isopods, 77, 78, 80, 82, 84, 85, 88, 89, 97, 136, 138, 147, 154, 155
isopterans, 85, 94, 151
jaguar, 17, 126
Jungle Frog, Smoky, **91–92**
Keelback, Ebony, **180–181**
Kinosternidae, 107, **110–113**
Kinosternon
    angustipons, 110, **111,** 112
    leucostomum, 110, 111, **112–113**
La Loma Caecilian, 26, **241**
Lachesis, 164
    melanocephala, 169
    muta, 169
    stenophrys, 17, 18, 164, 166, **168–169**
Lampropeltis triangulum, 173, 190, **195–196,** 210, 213, 220, 228, 230
land crabs, 159
laurel, 134, 135, 137
leaf frogs
    Gray-eyed Leaf Frog, 15, **47–48**
    Parachuting Red-eyed Leaf Frog, 49, **50–51**
    Red-eyed Leaf Frog, **49–50,** 244
    Spurrell's Leaf Frog, 50, **244**
Leaf-breeding Rainfrog, 76, **77–78**
leeches, 205
Leimadophis epinephalus, 202
Leiolopisma cherrei, 154
Lemur Anole, **136–137,** 246
lengua del Diablo, 9
Leopard Frog, 100–101

Lepidoblepharis xanthostigma, 144, 145, **146–147,** 148, 222
Lepidophyma flavimaculatum, **150–151,** 230
lepidopterans, 78, 85, 95, 137, 139
Leptodactylidae, 38, **73–92, 242–243**
Leptodactylus, 73
    melanonotus, 75, **90–91**
    pentadactylus, 75, **91–92**
    poecilochilus, 91, **243–244**
Leptodeira, 207
    annulata, 179, **196–197**
    septentrionalis, 50, 179, **197–198,** 248
Leptophis, 208
    ahaetulla, 174, **198–199,** 200
    depressirostris, 174, **199–200,** 201
    nebulosus, 173, **200–201**
Lichen Anole, **140–141**
Lichen-colored Slugeater, 165, **215**
Liophis epinephalus, 176, **201–202**
Litter Gecko, 145, **146–147,** 148, 222
Litter Skink, **153–154,** 157, 203
Litter Toad, **41–42**
Litterfrog, Bransford's, **76–77,** 84, 155, 185, 188, 200, 206, 212
littersnakes
    Long-tailed Littersnake, 212, 218, **222–223,** 225
    Orange-bellied Littersnake, 212, 218, 219, 222, 223, **224–225**
    Pink-bellied Littersnake, 183, **211–212,** 218, 219, 222, 223, 225
    Rugose Littersnake, **205–206**
    Short-tailed Littersnake, 16
Long-tailed Littersnake, 212, 218, **222–223,** 225
Long-tailed Worm Salamander, 30, **31–32**
Lower-montane Green Racer, **188–189**
Lowland Forest Racer, 184, **185**
Lowland Rainfrog, **87,** 90
Luehea seemannii, 10
Mabuya unimarginata, **152–153,** 157
mammals
    bats, 35, 123, 161, 211
    deer, 161, 233

Nine-banded Armadillo, 182
White-nosed Coati, 17, 161
Mammalia
   Agouti paca, 161
   Coendou rothschildi, 161
   Dasypus novemcinctus, 182
   Didelphis, 161
   Felis pardalis, 161
   Heteromys desmarestianus, 170
Many-banded Coralsnake, 190, 213,
   220, 228, **229–231**
maquenque, 9
Marine Toad, **42–43,** 44
Masked Treefrog, **60–61**
Mastigodryas melanolomus, 178, 179,
   181, 184, 185, **202–203**
mata gente, 9
Melastomataceae, 9
Microhylidae, 36, **93–94**
Micrurus
   alleni, 190, 195, 213, 220, **227,** 229
   mipartitus, 195, 224, 227, **228**
   nigrocinctus, 190, 213, 220, 227, 228,
   229
   nigrofasciata, 229
Milksnake, Tropical, 190, **195–196,**
   210, 213, 220, 228, 230
mongoose, 161
mosses, 28, 35
moths, 15, 149
mud turtles
   Narrow-bridged Mud Turtle, **111,**
   112
   White-lipped Mud Turtle, 111,
   112–113
Myrica splendens, 10
Narrow-bridged Mud Turtle, **111,** 112
Nasua narica, 17, 161
Neotropical Chameleon, **141–142**
Nicaraguan water tree, 10
Nine-banded Armadillo, 182
Ninia
   maculata, 176, **203–204**
   sebae, 175, **204–205**
Noble's Rainfrog, 85, **86**
Norops
   biporcatus, 129, **131–132**
   capito, 129, **132–133**

   carpenteri, 129, **133–134**
   cupreus, 11
   humilis, 129, **135–136**
   lemurinus, 129, **136–137,** 246
   limifrons, 129, **137–138**
   lionotus, 139
   oxylophus, 129, **139**
   pentaprion, 129, **140–141**
Northern Cat-eyed Snake, 50,
   197–198, 248
Northern Masked Rainfrog, **85–86,**
   205, 224, 243
Nothopsis rugosus, 176, 205–206
ocelot, 161
Oedipina
   cyclocauda, 29, 30, **31–32**
   gracilis, 29, 30, **31–32**
   pseudouniformis, 32
   uniformis, 32
oligochaetes, 26, 94
Olive Snouted Treefrog, 55, 57, **58–59,**
   198, 201
Ololygon, 58, 59
opossum, 161
Orange-bellied Crowned Snake,
   218–219, 221, 225
Orange-bellied Littersnake, 212, 218,
   219, 222, 223, **224–225**
Orange-bellied Swamp Snake, 178,
   220–221
Orange-billed Sparrow, 161
oranges, 11
orchids, 35,
Ornate Slider, 16, 116, 246
orthopterans, 66, 77, 78, 80, 82, 84, 89,
   94, 97, 137, 138, 147, 153, 155
Otoba novogranatensis, 10
Oxybelis, 201
   aeneus, 178, **206–207**
   brevirostris, 174, 198–199, **207–208**
   fulgidus, 173, 199, **208–209**
Oxyrhopus
   petola, 210
   petolarius, 171, 190, 195, **209–210,**
   213, 220, 224, 228, 229, 230
Pachira aquatica, 10
palma conga, 9
Panicum grande, 10

Parachuting Red-eyed Leaf Frog, 49, **50–51**
parrotsnakes
  Green Parrotsnake, **198–199,** 200
  Satiny Parrotsnake, **199–200,** 201
  Striped Parrotsnake, **200–201**
pejiballe, 134, 135, 137
Pentaclethra macroloba, 9
Phyllobates lugubris, 95, **97–98,** 242, 243
Phrynohyas venulosa, 55
Pink-bellied Littersnake, 183, **211–212,** 218, 219, 222, 223, 225
Pink-headed Blindsnake, **247**
Pipa, 17
Piperaceae, 9
Pithecellobium
  elegans, 9
  longifolium, 10
plants
  ajillo, 9, 10
  almendro, 9
  anturto blanco, 10
  bananas, 11
  bola de oro, 10, 247
  bromeliads, 28, 31, 35, 62, 82, 97, 193, 199, 200
  cacao, 11, 67, 96, 133, 134, 135, 137, 146, 148, 153, 184, 218, 223
  cafecillo, 10
  chilamate, 10
  chocolate, 11
  clavillo, 10
  cola de gallo, 56
  consuelo de mujer, 9
  copal, 9
  Costa Rica feather palm, 9
  Costa Rican fruta de pava, 10
  coyolillo, 10
  ferns, 35, 115
  guácimo colorado, 10
  huevos de burro, 10
  laurel, 134, 135, 137
  lengua del diablo, 9
  maquenque, 9
  mata gente, 9
  mosses, 28, 35
  Nicaraguan water tree, 10
  oranges, 11
  orchids, 35
  palma conga, 9
  pejiballe, 134, 135, 137
  sangregado, 10
  sota caballo, 10
  tabacón, 10
  turrú colorado, 10
  wild banana, 123, 166
  yaya, 9
  zacate grande, 10
Plethodontidae, **29–32, 241–242**
Pliocercus euryzona, 224
Pocket Mouse, Spiny, 170
poison frogs
  Green and Black Poison Frog, **95–96**
  Strawberry Poison Frog, 12, **96–97**
Polychrotidae, 120, **128–142, 246**
Polychrus gutturosus, 129, **141–142**
ponerine ants, 94
porcupines, 161
Porthidium nasutum, 164, **169–170**
poultry, 161
Powdered Glassfrog, 67, **71–72**
Proechimys, 168
Protium pittieri, 9
Pseustes poecilonotus, 174, 188, **210–211**
Psychotria chagrensis, 10
Pterocarpus officinalis, 10
Puerto Rican Crested Anole, 137, **246**
Pug-nosed Anole, **132–133**
Purple Caecilian, **25–26,** 241
Pygmy Rainfrog, 79, 81, 82, **88**
rabbits, 161
racers
  Brown Forest Racer, **184,** 185, 203
  Lowland Forest Racer, 184, **185,** 203
  Lower-montane Green Racer, **188–189**
  Salmon-bellied Racer, 181, 184, 185, **202–203**
  Speckled Racer, **187–188**
Rainbow Boa, **247**
Rainforest Celestus, 152, 153, **156–157**
rainfrogs

Broad-headed Rainfrog, **84,** 224, 226, 242

Clay-colored Rainfrog, **78–79,** 81, 88

Coral-spotted Rainfrog, **75–76**

Fitzinger's Rainfrog, 80, **82–83,** 86, 89, 188, 198

Gaige's Rainfrog, 98, **242–243**

Golden-groined Rainfrog, 76, 79, **80–81,** 88

Leaf-breeding Rainfrog, 76, **77–78**

Lowland Rainfrog, **87,** 90

Noble's Rainfrog, 85, **86**

Northern Masked Rainfrog, **85–86,** 205, 224, 243

Pygmy Rainfrog, 79, 81, 82, **88**

Slim-fingered Rainfrog, **79–80,** 83, 86, 89

Southern Masked Rainfrog, 86, **243**

Talamancan Rainfrog, 80, 83, 86, **89**

Warty Rainfrog, 84, **242**

Rana
  palmipes, 100, 101
  pipiens, 101
  taylori, 101, **245**
  vaillanti, 91, **99–101,** 197, 199, 203
  warszewitschii, 99, **101–102,** 170

Ranidae, 38, **99–102, 245**

Ratsnake, Tiger, 181, **216–217**

Red Coffeesnake, **204–205**

Red-eyed Leaf Frog, **49–50,** 244

Reptilia, **103–237**

Reticulated Crowned Snake, **217–218,** 219, 221, 225

Reticulated Glassfrog, **72**

Reticulated Sheepfrog, **93–94,** 181, 188

Rhadinaea
  decipiens, 223
  decorata, 177, 183, **211–212,** 218, 219, 222, 223, 225
  guentheri, 225

Rhinobothryum bovallii, 230–231, **248**

Rhinoclemmys
  annulata, 113, **114–115**
  funerea, 113, 114, **115–116**

Ridge-headed Salamander, **30–31,** 242

Ridge-nosed Snake, **179–180,** 189, 191

Ringed Slugeater, 186, **214–215,** 216

Round-tailed Worm Salamander, 30, **31–32**

Rubiaceae, 9

Rugose Littersnake, **205–206**

Ruiz's guabo, 10

Ruthven's Earthsnake, 16

salamanders
  Alvarado's Salamander, 11, 16, 31, **241–242**
  Ridge-headed Salamander, **30–31,** 242
  Striated Salamander, **242**
  See also worm salamanders

Salmon-bellied Racer, 181, 184, 185, **202–203**

San Carlos Treefrog, 52, 53, **54–55,** 58

sangregado, 10

Satiny Parrotsnake, **199–200,** 201

Scaphiodontophis
  annulatus, 172, 176, 190, 195, **212–213,** 220, 228, 230
  venustissiumus, 214

Scarlet-webbed Treefrog, **55–56**

Scinax
  boulengeri, 46, **57–58,** 59
  elaeochroa, 46, 55, 57, **58–59,** 198, 201

Scincidae, 120, **151–154**

Scorpion-eater, Degenhardt's, 16, 198, **248–249**

Seep Snake, Tropical, **192**

Sheepfrog, Reticulated, **93–94,** 181, 188

Short-nosed Vinesnake, 198–199, **207–208**

Short-tailed Littersnake, 16

shrubs, 8, 9, 11, 28, 31, 66, 76, 78, 79, 115, 123, 125, 185, 193, 211, 248

Sibon, 216
  annulatus, 178, 186, **214–215,** 216
  longifrenis, 165, 174, **215**
  nebulatus, 176, 204, **216**

skinks, 213
  Bronze-backed Climbing Skink, **152–153,** 157
  Litter Skink, **153–154,** 157, 203

Skinkeater, 190, 195, **212–213,** 220, 228, 230

Slender Anole, **137–138**

Slider, Ornate, 16, 116, **246**
Slim-fingered Rainfrog, **79–80,** 83, 86, 89
slugeaters
    Cloudy Slugeater, 204, **216**
    Lichen-colored Slugeater, 165, **215**
    Ringed Slugeater, 186, **214–215,** 216
slugs, 204, 205, 215, 216
Smilisca, 54, 61, 62, 63
    baudinii, 47, **59–60,** 181, 199
    phaeota, 47, **60–61**
    puma, 47, **62**
    sordida, 47, **63–64**
Smoky Jungle Frog, **91–92**
Snaileater, Bicolored, **186**
snakes
    Bird-eating Snake, 188–189, **210–211**
    Brown Debris Snake, **183–184**
    Calico Snake, 190, 195, **209–210,** 213, 220, 224, 228, 229, 230
    Faded Dwarf Snake, 218, 219, **221–222,** 223, 225
    Fire-bellied Snake, **201–202**
    Halloween Snake, 171, 210, **223–224,** 229
    Northern Cat-eyed Snake, 50, **197–198,** 248
    Orange-bellied Crowned Snake, **218–219,** 221, 225
    Orange-bellied Swamp Snake, 178, **220–221**
    Reticulated Crowned Snake, **217–218,** 219, 221, 225
    Ridge-nosed Snake, **179–180,** 189, 191
    Southern Cat-eyed Snake, **196–197**
    Tricolored Crowned Snake, 190, 195, 210, 213, **219–220,** 228, 230
    Tropical Seep Snake, **192**
    Viquez's Dwarf Snake, 212, 222, **249**
    White-headed Snake, 179, 180, **189,** 191
    See also Blindsnake, Pink-headed; Coffeesnake, Red; coralsnakes; earthsnakes; littersnakes; Milksnake, Tropical; parrotsnakes; Ratsnake, Tiger; vinesnakes

Snapping Turtle, **109–110**
snouted treefrogs
    Boulenger's Snouted Treefrog, **57–58,** 59
    Olive Snouted Treefrog, 55, 57, **58–59,** 198, 201
Socrotea exorrhiza, 9
sota caballo, 10
Southern Cat-eyed Snake, **196–197**
Southern Masked Rainfrog, 86, **243**
Sparrow, Orange-billed, 161
Spathiphyllum freidrichsthallii, 10
Speckled Racer, **187–188**
Spectacled Caiman, 17, **234–235,** 236
Sphaerodactylus
    homolepis, 143, 146, **147–148**
    millepunctatus, 143, 145, 146, 147, **148**
Sphenomorphus, 213
    cherrei, 152, **153–154,** 157, 203
Spilotes pullatus, 176, 181, **216–217**
Spined Glassfrog, 66, **69**
Spiny Pocket Mouse, 170
Spiny Rat, 168
Spiny-tailed Iguana, 161
Spotted Dwarf Gecko, 146, **147–148**
Spotted Woodsnake, **203–204**
Spurrell's Leaf Frog, 50, **244**
Squamata, 105, **117–231**
Stenorrhina degenhardtii, 16, 198, **248–249**
Strawberry Poison Frog, 12, **96–97**
Stream Anole, **139**
Streptoprocne zonaris, 211
Striated Salamander, **242**
Striped Basilisk, 18, **124–125**
Striped Dart-poison Frog, **97–98,** 242, 243
Striped Parrotsnake, **200–201**
Swamp Snake, Orange-bellied, 178, **220–221**
Swamp Treefrog, **53–54**
Swift, White-collared, 211
tabacón, 11
Talamancan Galliwasp, **157–158**
Talamancan Rainfrog, 80, 83, 86, **89**
Tanager, Blue-gray, 161
Tantilla, 212
    annulata, 220

melanocephala, 219
reticulata, 177, **217–218,** 219, 221, 225
ruficeps, 177, **218–219,** 221, 225
supracincta, 173, 190, 195, 210, 213, **219–220,** 228, 230
Tawny Treefrog, **62**
Taylor's Leopard Frog, 100–101, **245**
Teiidae, 120, **154–155, 247**
teleost fishes, 27
Terciopelo, **166–167,** 168
Testudines, 105, **106–116, 246**
Thecadactylus rapicauda, 143, **149–150**
Theobroma cacao, 67, 96, 133, 134, 135, 137, 146, 148, 153, 184, 218, 223
Thraupis episcopus, 161
ticks, 115, 116, 135
Tiger Ratsnake, 181, **216–217**
toads
　Green Climbing Toad, **40–41,** 44
　Litter Toad, **41–42**
　Marine Toad, **42–43,** 44
　Wet Forest Toad, 41, **44–45**
Trachemys ornata, 16, 116, **246**
Tree Boa, Annulated, **161–162**
Tree Coral, False, 230–231, **248**
Tree Porcupine, 161
treefrogs
　Boulenger's Snouted Treefrog, **57–58,** 59
　Common Mexican Treefrog, **59–60,** 181, 199
　Drab Treefrog, **63–64**
　Harlequin Treefrog, **52–53,** 54, 55, 185
　Highland Fringe-limbed Treefrog, 61, **244**
　Masked Treefrog, **60–61**
　Olive Snouted Treefrog, 55, 57, **58–59,** 198, 201
　San Carlos Treefrog, 52, 53, **54–55,** 58
　Scarlet-webbed Treefrog, **55–56**
　Swamp Treefrog, **53–54**
　Tawny Treefrog, **62**
Tretanorhinus nigroluteus, 178, **220–221**

Tricolored Crowned Snake, 190, 195, 210, 213, **219–220,** 228, 230
Trimetopon
　pliolepis, 177, 218, 219, **221–222,** 223, 225
　viquezi, 212, 222, **249**
Tropical Milksnake, 190, **195–196,** 210, 213, 220, 228, 230
Tropical Seep Snake, **192**
Turbo White-lipped Foamfrog, 91, **243–244**
Turnip-tailed Gecko, **149–150**
turrú colorado, 10
turtles
　Black Wood Turtle, 114, **115–116**
　Brown Wood Turtle, **114–115**
　Narrow-bridged Mud Turtle, **111,** 112
　Snapping Turtle, **109–110**
　White-lipped Mud Turtle, 111, **112–113**
Ungaliophiidae, 121, **162–163**
Ungaliophis panamensis, **163**
Unonopsis pittieri, 9
Urotheca
　decipiens, 177, 212, 218, **222–223,** 225
　euryzona, 171, 210, **223–224,** 229
　guentheri, 177, 212, 218, 219, 222, 223, **224–225**
　pachyura, 16
Vaillant's Frog, 91, **99–101,** 197, 199, 203
vines, 8, 11, 49, 115, 207
vinesnakes
　Brown Blunt-headed Vinesnake, **193–194,** 195
　Brown Vinesnake, **206–207**
　Green Vinesnake, 199, **208–209**
　Short-nosed Vinesnake, 198–199, **207–208**
　Yellow Blunt-headed Vinesnake, 193, **194–195**
Viperidae, 120, 162, **164–170**
Viquez's Dwarf Snake, 212, 222, **249**
Warscewiczia coccinea, 9
Warty Rainfrog, 84, **242**
water tree, Nicaraguan, 10

Wedge-billed Woodcreeper, 170
Welfia georgii, 9
Wet Forest Toad, 41, **44–45**
Whiptail, Central American, 15,
    154–155, 203, 207, 209, 210, 247
White-collared Swift, 211
White-headed Snake, 179, 180, **189,**
    191
White-lipped Mud Turtle, 111,
    112–113
White-nosed Coati, 17, 161
wild banana, 123, 166
wood turtles
    Black Wood Turtle, 114, **115–116**
    Brown Wood Turtle, **114–115**
Woodcreeper, Wedge-billed, 170
Woodsnake, Spotted, **203–204**

worm salamanders
    Long-tailed Worm Salamander, 30,
        31–32
    Round-tailed Worm Salamander,
        30, **31–32**
Xantusiidae, 120, **150–151**
Xenodon rabdocephalus, 166, 178,
        225–226
yaya, 9
Yellow Blunt-headed Vinesnake, 193,
        194–195
Yellow-flecked Glassfrog, **67,** 70, 71
Yellow-headed Gecko, 16, **145–146,**
        148
Yellow-tailed Dwarf Gecko, 145, 146,
        147, **148**
zacate grande, 10

Original Design:	Barbara Jellow
Design Enhancements:	Beth Hansen
Design Development:	Jane Tenenbaum
Cartographer:	Bill Nelson
Composition:	Impressions Book and Journal Services, Inc.
Text:	9/10.5 Minion
Display:	Franklin Gothic Book and Demi
Printer and Binder:	Everbest Printing Company